T0214773

The Common Worlds of Children and Animals

The lives and futures of children and animals are linked to environmental challenges associated with the Anthropocene and the acceleration of human-caused extinctions. This book sparks a fascinating interdisciplinary conversation about child–animal relations, calling for a radical shift in how we understand our relationship with other animals and our place in the world.

It addresses issues of interspecies and intergenerational environmental justice through examining the entanglement of children's and animal's lives and common worlds. It explores everyday encounters and unfolding relations between children and urban wildlife. Inspired by feminist environmental philosophies and indigenous cosmologies, the book poses a new relational ethics based upon the small achievements of child–animal interactions. It also provides an analysis of animal narratives in children's popular culture. It traces the geo-historical trajectories and convergences of these narratives and of the lives of children and animals in settler-colonised lands.

This innovative book brings together the fields of more-than-human geography, childhood studies, multispecies studies, and the environmental humanities. It will be of interest to students and scholars who are reconsidering the ethics of child–animal relations from a fresh perspective.

Affrica Taylor is an adjunct associate professor at the University of Canberra, Australia. Her background in cultural geography and Indigenous Australian education have shaped her abiding interest in the relations between people, place, and other species in settler colonial societies, and in the need to decolonise these relations. She explores these themes in her books, *Reconfiguring the Natures of Childhood* and *Unsettling the Colonial Places and Spaces of Early Childhood Education*.

Veronica Pacini-Ketchabaw is a professor of early childhood education at Western University in Canada. She is committed to tracing the common world relations of children with places, materials, and other species. Her books *Encounters with Materials in Early Childhood Education, Unsettling the Colonial Places and Spaces of Early Childhood Education, Journeys: Reconceptualizing Early Childhood Practices*, and *Flows, Rhythms, and Intensities of Early Childhood Education Curriculum* explore these relations.

Routledge Spaces of Childhood and Youth Series
Edited by Peter Kraftl and John Horton

The *Routledge Spaces of Childhood and Youth Series* provides a forum for original, interdisciplinary and cutting-edge research to explore the lives of children and young people across the social sciences and humanities. Reflecting contemporary interest in spatial processes and metaphors across several disciplines, titles within the series explore a range of ways in which concepts such as space, place, spatiality, geographical scale, movement/mobilities, networks and flows may be deployed in childhood and youth scholarship. This series provides a forum for new theoretical, empirical and methodological perspectives and ground-breaking research that reflects the wealth of research currently being undertaken. Proposals that are cross-disciplinary, comparative and/or use mixed or creative methods are particularly welcomed, as are proposals that offer critical perspectives on the role of spatial theory in understanding children and young people's lives. The series is aimed at upper-level undergraduates, research students and academics, appealing to geographers as well as the broader social sciences, arts and humanities.

For more information about this series, please visit: www.routledge.com/Routledge-Spaces-of-Childhood-and-Youth-Series/book-series/RSCYS

The Common Worlds of Children and Animals

Relational Ethics for Entangled Lives

Affrica Taylor and Veronica Pacini-Ketchabaw

Routledge
Taylor & Francis Group

LONDON AND NEW YORK

First published 2019
by Routledge
2 Park Square, Milton Park, Abingdon, Oxon OX14 4RN

and by Routledge
52 Vanderbilt Avenue, New York, NY 10017

First issued in paperback 2020

Routledge is an imprint of the Taylor & Francis Group, an informa business

British Library Cataloguing-in-Publication Data
A catalogue record for this book is available from the British Library

Library of Congress Cataloging-in-Publication Data
A catalog record for this book has been requested

ISBN 13: 978-0-36-758513-6 (pbk)
ISBN 13: 978-1-138-94759-7 (hbk)

Typeset in Times New Roman
by Apex CoVantage, LLC

We dedicate this book to Deborah Bird Rose. We would like to thank her for championing ethical considerations of multispecies relations in the face of extinction and for her generous and gracious support of our work with children and animals.

Contents

Figures

Acknowledgements

Affrica

I would like to acknowledge and thank all the children, teachers, educators, colleagues, and resident wildlife (although they didn't know it) who participated in my evolving common worlds multispecies ethnographic research across two Canberra-based university campuses from 2013 to 2017.

From 2013 to 2015, I conducted a multispecies ethnographic research project called 'Common Worlds Pedagogies', based at the Wiradjuri Early Childhood Centre and Preschool at University of Canberra. Adam Duncan, a Koori teacher at Wiradjuri, and I worked as a close team during this time. We took the children for weekly walks on the University of Canberra campus grounds; observed and reflected upon the unfolding relations with kangaroos, ants, birds, fungi, spiders, bugs, tadpoles, and other resident wildlife; read together; attended an international symposium together; and wrote articles about our settler-indigenous collaboration. Adam's Aboriginal perspectives on child–animal common worlds and pedagogies greatly enriched our project and fed directly into the sections on child-kangaroo encounters and thinking with ants.

From 2016 to 2017, I joined forces with Tonya Rooney from Australian Catholic University and embarked upon a new iteration of this multispecies ethnographic research called 'Walking with Wildlife in Wild Weather Times'. This project was designed to further explore the possibilities of more-than-human environmental pedagogies in the face of anthropogenic climate change and species extinctions. It was conducted on the campus of the Australian National University (ANU) with staff and children from the ANU University Preschool and Childcare Centre. I am indebted to Tonya for the weathering perspectives she brought to the project, our post-walk debriefing conversations, and of course for all the times we spent walking/staggering with children and wildlife in the bitterly cold and windy Canberra winters and the scorching Canberra summer heat waves.

Both projects received ethics approval from the University of Canberra Ethics Committee. I would also like to acknowledge the support of the parents and families of the children taking part in the research, some of whom occasionally joined us on the walks.

I would also like to acknowledge the nourishing and sustaining support of my colleagues in the Common Worlds Research Collective: Veronica Pacini-Ketchabaw, Mindy Blaise, Tonya Rooney, Catherine Hamm, Margaret Somerville, Karen Malone, Suzanne Gannon, Iris Duhn, Fikile Nxumalo, Nicole Land, Carol Rowan, Narda Nelson, Paulina Rautio, Peter Kraft, Susannah Clement, and Karin Murris, to name just a few . . . as well as other significant fellow travellers like Lesley Instone, Katherine Gibson, and Deborah Bird Rose. Without the multiple meetings up at conferences all over the world, cumulative serious mind-stretching conversations, irreverent hilarities, and crazy escapades (especially during bush salon events at Wee Jasper), this book would not have shaped up as it is.

Veronica

I would like to acknowledge and thank all the children, educators, and creatures who generously gave of their time, knowingly or unknowingly, to the common worlds multispecies ethnographic research that took place in early childhood education centres in Victoria and Vancouver, British Columbia, from 2011 to 2016. Their dedication and insightful provocations made the research super interesting.

This research was supported by the Social Sciences and Humanities Research Council of Canada.

I would also like to acknowledge the many other people who provided a supportive and rigorous scholarly environment throughout the writing of this book: Affrica Taylor, Fikile Nxumalo, Narda Nelson, Myra Hird, Mindy Blaise, Nicole Land, Carol Rowan, Denise Hodgins, Kathleen Kummen, Vanessa Clark, Cristina Delgado, Sylvia Kind, Laurie Kocher, Catherine Hamm, and many others. The astute conversations that they willingly had with me made my contributions to this book better.

I finally want to thank Rob and Jacob who always accompany me on the ride. Their support and love sustain me.

Introduction

> We must connect the question of the common world to the question of the common good.
>
> (Latour 2004, p. 98)

Common worlds and the common good

Throughout this book, we hold two questions in productive tension. The first is the onto-epistemological question: Who and what constitutes the common worlds that children and animals co-inherit and co-inhabit? The second is the ethical question: How might we approach these common worlds in ways that attend to the common good of all its constituents?

We do not believe that such questions can be satisfactorily answered by applying the humanist logic that pervades the mainstream social sciences, including the mainstream of the disciplinary fields that we straddle. The majority of scholars in childhood studies and children's geographies, for instance, still maintain an overriding commitment to exclusively human(ist) notions of social inclusion and social justice. This means that they predominantly uphold child-centred approaches, champion children's agency and rights, and contextualise child development within exclusively sociocultural contexts.[1] Prompted by a series of epistemological turns that navigate around the human-centrisms of social science scholarship – such as the 'animal turn', the 'feminist material turn', and the 'ontological turn' – we move from a different set of premises. Although we also champion inclusion and justice, we pursue it within more-than-human framings.[2]

We acknowledge that children grow up in geographically, biologically, and culturally diverse communities – not just in human societies and not just according to the laws of one essentialised concept of 'nature'. Moreover, we believe that all of this heterogeneity matters. Borrowing from Latour (2004, 2011, 2014), we refer to these pluralist human and nonhuman communities as 'common worlds'. We situate our studies of child–animal relations within these common worlds, paying close attention to the interplay of their geo-historical, bio, and cultural specificities. We are particularly interested in the ways in which children's and animals' lives are entangled within these common worlds. This draws us to follow child–animal relations, not individual children, as is customary in childhood studies and geographies, and to trace the trajectories of these entangled relations – materially, semiotically, and ecologically.

Rather than regarding these common worlds as a stage on which all-important human lives, activities, and agencies take place, we see them as emerging from, and continuing to be shaped and reshaped by, the collective relations and interactions of human and nonhuman beings, entities, and forces. In other words, we see common worlds as permanently evolving and never closed – always in the process of being made and remade through the 'dance of encounters' between their human and nonhuman constituents (Haraway 2008, p. 4). Common worlds are more like a verb than a noun, more like the active processes of commoning or 'worlding', a term that Donna Haraway often uses and defines as 'making worlds together' (see Haraway 2016).

It stands to bear, then, that agency within common worlds cannot be exclusively human and is never the sole outcome of individual intent. We see it as is dispersed, relational, collective, and interactive – simultaneously productive of and contingent upon the specificities of a diversity of common worlds and their constituents. So instead of promoting children's autonomous and individual agency, we refocus upon children's and animals' collective agency. We do this by tracing the ways in which particular kinds of entangled child–animal relations produce certain ways of being within specific common world contexts and, in the process, produce distinctive kinds of worlds.

By including all manner of lifeforms, forces, and entities – everything that makes up a world – it is clear that a common worlds framework extends a commitment to inclusion beyond human diversity. In the same way, thinking through the prism of common worlds requires that we also extend the notion of justice beyond the human. Justice cannot only be for us if our lives are already inextricably entangled with the full range of heterogeneous others in our common worlds. This kind of collective logic goes some way to ensuring that the common good remains the central concern of the common world.

The entangled lives of children and animals

To answer our first framing question – 'who and what constitutes the common worlds that children and animals co-inherit and co-inhabit?' – we trace the ways in which the lives of children and animals are materially, semiotically, and ecologically entangled and mutually constitutive.

The material entanglement of children's and animals' lives is self-evident in the case of children who live and grow up with pets or spend significant amounts of time with other kinds of companion species, such as riding horses, or with village, farm, or herd animals, such as chickens, pigs, sheep, and goats. In most parts of the world, this means a very significant proportion of children. These material child–animal relations are embodied and emplaced – produced through everyday physical proximities, routines, and familiarities, as well as varied exchanges of companionship, care, food production and supply, learning, exercise, and labour.

They also attract a large degree of social endorsement. Within western romantic traditions, there is widespread belief that children have a natural affinity with animals. Within the biological sciences, this belief has been promoted by E.O. Wilson's (1984) 'biophilia' hypothesis, which draws upon evolution to assert that we are born with an innate attraction to other species because of our common ancestry. Children's relationships with pets and other companion species are often seen as beneficial to their growth and development. For instance, within the framings of developmental psychology, children's loving and caring relations with such animals are valued as enabling of positive social relations (Kahn and Kellert 2002; Myers 2007). They are also embraced as potentially therapeutic and as an aid to children's learning, particularly within some branches of special education. For instance, 'animal assisted learning' programs are based upon the premise that children with physical disabilities and special behavioural needs can gain confidence and build skills through interacting with suitable domestic and farm animals. This kind of environment is seen to optimise the conditions for the children's learning, because their interactions with the animals are emotionally stimulating and rewarding, and their stress is reduced because the animals are nonjudgemental (McCardle *et al.* 2011; Ross 2011).

Beyond the physical and emotional intimacies of domesticated human-animal relations, children's lives are also materially entangled with those of wild animals, although it is perhaps not quite so obvious exactly how this transpires. Although uncommon, one clear example is children who grow up in remote hunting and fishing communities. In such communities, wildlife interactions are a fundamental part of everyday life, vital to survival, and explicitly framed by the mortal relations of hunters and prey. Through direct observation and embodied experience, children are taught how to hunt, kill, and butcher wild animals to eat. In less visceral ways, the lives of urban children are also materially entwined with wild animals. This is simply because they coexist. From birds and bees through to squirrels, raccoons, and possums, all kinds of wildlife inhabit urban spaces. Moreover, because of the scale of habitat destruction and the cascading effects of other mitigating human-induced changes, such as global warming, the migration of displaced wildlife into urban areas is an increasing trend (see Lorimer 2015). Their survival depends upon access to food and shelter in human-purposed environments, many of which are frequented by children – for instance, urban parks and school and childcare centre yards, where food and shelter abounds.

In contrast to the widespread fond regard bestowed on domesticated animals, the majority of wild animals that reside and feed in urban environments are seen as unwelcome interlopers, as out of place in towns and cities. Wild animals are by definition beyond human containment or control. So, as they move freely within human-purposed domains, confounding the rules and regulations of social organisation, they are often branded a public nuisance or a health and safety threat, particularly to vulnerable people like children.

For all of these reasons, we are particular interested in the material encounters, interactions, and entanglements of children and wild animals in urban settings. Although increasingly commonplace, real-life urban child-wildlife relations have not attracted much attention within childhood studies and children's geographies. This is assumedly because they do not carry the same social endorsement that is attributed to relations between children and domesticated or farm animals. It seems that relations with wildlife have been either overlooked or undervalued because it is hard to see how they might advantage children. However, beyond the more traditional child-centred perspective, the increasingly entangled daily lives of children and wild animals in towns and cities have broader ecological, peda-gogical, and ethical implications, which we explore in detail in the forthcoming chapters. By drawing upon our own ethnographic observations of child-wildlife encounters and interactions, we pursue the question of how we might live well with others that are so radically different to us. Or, to elaborate upon Latour's question cited earlier (2004, p. 98), how we might attend to the common good in our heterogeneous multispecies common worlds in the face of incommensurable differences.

Actual contact between living children and animals is never outside of a dis-cursive/semiotic context. There is always a cultural inheritance that precedes the encounter and shapes it in some sort of way. In the case of child–animal entangle-ments, this often comes from the animal narratives that proliferate in children's fiction and popular culture.

We are not the first to note that childhood is a key site in which human-animal relations are formed at the intersections of real-life physical encounters and cul-tural representations. With a focus on western societies, Matthew Cole and Kate Stewart (2014) argue that human-animal relations are culturally constructed in childhood through mediated child–animal encounters and media representations. Gail Melson (2005) similarly points out that North American children's lives are saturated with animal presences, both material and symbolic. She points to the high proportion of children who grow up with pets in the USA and to the plethora of animal motifs and commercial products that pervade these children's lives.

The material/semiotic entanglement of children's and animals' lives is not lim-ited to Euro-western contexts. All over the world, children have some kinds of face-to-face encounters with some kinds of animals. All cultures pass on animal stories to children – whether in the form of creation stories, allegorical fables or moral tales, cute anthropomorphic stories, light-hearted comedies, parodies, wild colonial tales, or serious ecological narratives. And all cultures produce pictorial or other kinds of visual representations of animals. In contemporary consumer societies, animal motifs adorn products made specifically for a children's market, and toys and all manner of other kinds of manufactured children's paraphernalia take the form of animals. The plethora of commercial animal representations on children's products often feature distorted, humanesque animal faces to increase their 'cute' appeal. This both plays upon and performatively reiterates the belief the children have some kind of 'natural' attraction to animals that predisposes them to a fondness for and affinity with animals.

However, there is never just one story. Children's animal stories can also be cautionary. Children's narratives that feature threatening and dangerous wild ani-mals are also part of the cultural imaginary of childhood. They predispose some children to fear certain animals, particularly 'exotic' wild animals that they are unlikely to actually encounter. So too do popular discourses about specific animals as 'pests', 'vermin', 'dirty', 'aggressive', 'invasive', 'creepy', etc. Children are not immune to the circulation of popularly held negative beliefs about certain animals which are transmitted through everyday conversations, via social media, on the television, and through the negative reactions that children observe displayed by the adults around them.

The complex semiotics of animal presences infuse children's lives, as do the accompanying values, attitudes, beliefs, and understandings about specific animals and our relationship to them. Their impact is unquestionable, but also very geo-historically and culturally situated and contingent, and, of course, inextricably entwined with children's material relations with living animals. In the situated studies that compose the various chapters of this book, we try and tease out the interwoven semiotic/material trajectories that are both constitutive of and reconstituted by everyday child–animal encounters.

We also attend to the ways that the micro-effects of these everyday child–animal encounters are part of the macro-politics of mortal ecological entanglements. An appreciation of the vital interdependence of all lifeforms and the precarity of our current ecological communities frame our interest in child–animal relations. It also addresses the 'so what?' question that could be posed about the significance of these relations.

Our interest in interspecies relations is premised on an understanding that all life on earth is enabled by the same geo-bio-chemical conditions of possibility and all species share a common ancestral trajectory. In other words, we are always and already primaeval kin. Not only do we share entangled life trajectories with other animals in a deep-time sense, but we are also bound together in the here and now. The lives of all species are conditional upon the lives and deaths of others, in both evolutionary and contemporaneous ecological terms. Deborah Bird Rose (2012, p. 139) refers to both the sequential and synchronistic modes of these intra- and interspecies entanglements as 'multispecies knots of ethical time', in which 'the great patterns of life, death, sustenance and renewal . . . intersect across species and generations to form flows of life-giving life'.

Full recognition of our mortal entanglement with other species leads to the concomitant recognition of mutual interspecies vulnerability. Moreover, once recognised, this knowledge carries a considerable ethical responsibility, not only to find ways to secure the ecological futures of our own children, but the future generations of all other species, with whom our fates and futures are irrevocably bound.

Common world ethics

The second question that drives our endeavours in this book – 'how might we approach these common worlds in ways that attend to the common good of all its constituents?' – is an ethical one. It is motivated by our commitment to seeking intergenerational and multispecies environmental justice in ecologically precarious times, but not on our own. We do not consider ethics to be an exclusively human concern or capacity to 'do good' according to prescribed moral codes, and we are not interested in heroic, paternalistic stewardship. For us, common world ethics are ongoing relational practices involving human and more-than-human actors and situated within the ordinary interactions and exchanges of everyday life.

We are interested in what we might learn about common world ethics by observing what is already going on in the everyday exchanges between children and animals. Like Maria Puig de la Bellacasa (2010), we are hopeful that there may be something about the doings and the effects of our everyday relations with other living beings that offers new ethical possibilities. We believe that 'something' can be found in the small interspecies achievements of certain kinds of child–animal exchanges and interactions.

We stress that it is only certain kinds of child–animal exchanges and interactions that might illuminate how common world ethics are practised, because we do not want to idealise, romanticise, or valorise them all. They can be simply rehearsals of 'man over nature' through which children learn how to separate off from other animals in order to manipulate or dominate them. The kinds of exchanges we are interested in are ones in which children notice and seek connection with animals, where recognising difference in common being is the stimulus for mutual curiosity and respect between children and animals, and where the forces of mutual affect shape child–animal relations of care.

Common world methods

In line with these ethics, our common world methods are similarly relational, more-than-human, and situated. We deliberately sidestep the aforementioned child-centred research methods favoured within childhood studies and children's geographies. Instead, we choose methods that help us trace the discursive, fleshy, and mortal entanglements of children and animals in situated common worlds and to reflect upon the productive tensions of child–animal relations. The mix of multispecies ethnographic, narrative, and genealogical methods we assemble helps us to better understand how worlds are shaped and reshaped together. Or, to put it in another way, the mix of more-than-human methods we draw upon helps us to witness the ways in which worlding and commoning occurs.

In order to avoid the exclusionary conceits of the kind of intellectual work that performatively reiterates the superiority of 'rational man', we selectively engage with feminist more-than-human methods that refuse to divide the world into humans and the rest, and which circumvent the mind/body split and any form of Cartesian 'cogito ergo sum' (I think therefore I am) idolatry. Our focus is upon relations between 'minor' players. Children and animals are routinely relegated to the inferior/minor domain of 'instinct' and 'nature' precisely because they assumedly lack the capacity for rational thought.

From the marginal worlds of children and animals in which we hang out, Ursula Le Guin's (1989, 1996) feminist 'carrier bag theory of fiction' and Donna Haraway's (2004) derivative 'bag lady story-telling' method both hold great appeal.

Le Guin (1989) originally adopted the carrier bag as a metaphor for writing that holds things/stories that have been selectively gathered 'at the end of the world'. She later elaborates that these things/stories in the carrier bag are held in powerful relation to each other and to the reader (1996). In direct conversation with Le Guin, Haraway (2004) extrapolates that her own 'bag lady' narrative method is the practice of 'putting unexpected partners and irreducible details into a frayed, porous carrier bag' (Haraway 2004, p. 127). For Haraway, of course, these unexpected partners are never exclusively human. Neither are they particularly special. She looks for them on the mundane, messy grounds of everyday life. It is here, she says, that we can 'learn to be worldly from grappling with, rather than generalizing from the ordinary' (Haraway 2008, p. 3). Our common world methods are similarly grounded in the ordinary every day. In full sympathy with Le Guin's and Haraway's feminist intents, we eschew the heroic tales of major individuals on the big stage and seek out alternative, minor, but powerful polyphonic stories of multiple small players, quietly changing worlds together on the margins.[3]

Our carrier bag holds an array of common worlding methods. These include narrative vignettes from our multispecies ethnographic studies of children and urban wildlife in Australia and Canada; geo-historical tracings of the trajectories and convergences of animals, settlers, indigenous people within settler-colonised lands; semiotic/material analysis of anglophone children's animal literature and popular culture; and reconceptualisations of child–animal relations informed by affect-attuned, more-than-human geographies, decolonising and feminist environmental philosophies, and indigenous cosmologies.

The chapters

In Chapter 1, 'The common worlds of children and animals', we lay out the foundations for our thinking about the common worlds of children and animals. There are many influences, so this takes us across a lot of territory. We begin by acknowledging the scholars and ideas that have informed our common worlds conceptual framework, charting a series of transdisciplinary dialogues that have cumulatively and collectively produced an alternative conceptual 'tool bag' for thinking beyond the anthropocentric limits of humanist intellectual traditions. It is this tool bag that we put to use in later chapters.

Global/local articulations of geo-political, ecological, cultural, and ethical issues and concerns are bedrock to our thinking. We are always aware that global issues, particularly the planetary-scale ecological damage spearheaded by imperialism and intensified by extractive capitalism, manifest quite specifically in different local environs. In this first chapter, our task is to provide a big-picture framework that outlines how these global/local articulations are relevant to the entangled lives of children and animals. We do this by offering a general overview of the patterns of children's and animals' entangled ecological inheritances, fictions, lives, and futures. The chapters that follow are based upon situated studies. They allow us to detail how these global/planetary issues and concerns are played out in geo-culturally specific ways in the local lives of children and animals.

In Chapter 2, 'Children, kangaroos, and deer: an ethics of multispecies con-viviality', we begin to explore the ethical implication of children's cohabitations with urban wildlife. We look at child-kangaroo cohabitations in an Australian city and child-deer cohabitations in a Canadian city. Drawing from our multispecies ethnographies in these two places, we offer detailed narrative vignettes of child–animal encounters. We provide context for these narratives firstly by tracing the geo-political histories of Australian settler–kangaroo and Canadian settler–deer relationships, secondly by looking at how child-kangaroo and child-deer relations are portrayed in children's literature and popular culture, and thirdly by linking the presence of kangaroo and deer in urban areas to anthropogenic environmental changes. Our first set of reflections on these particular examples of urban child–animal cohabitation draw upon Thom van Dooren and Deborah Bird Rose's (2012) notion of a multispecies ethics of conviviality, achieved by paying careful attention to the stories of all urban inhabitants.

Chapter 3, 'Children, ants, and worms: an ethics of mutual vulnerability', is also based around narratives that are drawn from our respective multispecies eth-nographic studies. In these narratives, we describe a series of children's encoun-ters with ants in Australia and with worms in Canada and focus upon the effect and vulnerability of both children and mini-creatures in these encounters. We also dip into the scientific literature about ants and worms as soil-makers and earth shapers in order to reinforce how our lives on earth are made possible and sus-tained by these small creatures. Myra Hird's (2010) 'microontologies' and 'envi-ronmental ethics of vulnerability' (2013) help us to think about being and ethics beyond human prototype and to reflect upon the significance of these child-worm and child-ant encounters as examples of an ethics of mutual vulnerability.

Unlike the previous two chapters, Chapter 4, 'Children, Bilbies, and spirit bears: a decolonising ethics of ecological reconciliation', is entirely based upon our analysis of children's cultural texts. We look at the ways in which child read-ers and audiences are interpolated by these texts' eco-nationalist narratives about 'endangered' native animals that they are highly unlikely to ever encounter – bilbies in Australia and spirit bears in Canada. The settler colonial contexts of these eco-narratives are central to our analysis of them. Despite their honour-able intentions to protect endangered native species and indeed in some cases to endorse indigenous knowledges, we show how these narratives inadvertently reinscribe neocolonialism. We trace how the recurring tropes of settler invasion, destruction, and salvation within these children's texts ultimately displace or appropriate indigenous people and knowledges and then position children and young people as the 'natural' protectors or saviours of endangered native animals. Drawing upon Deborah Bird Rose's (2004, 2011) eco-philosophies, informed by her Aboriginal teachers in northern Australia, we conclude by considering how we might interrupt and augment such white-settler children's narratives with a decolonising ethics of ecological reconciliation.

In Chapter 5, 'Children, raccoons, and possums: an ethics of staying with the trouble', we return to recounting stories from our multispecies ethnographic stud-ies of children and urban wildlife. In this case, we focus upon the difficulties, tensions, and delights that occur when children cohabit in close quarters with raccoons in a Canadian early childhood centre and with possums in an Australian early childhood centre. Once again, we provide vignettes of actual encounters, discuss the cultural histories of possum-settler and raccoon-settler relations, and take a look at children's books that feature these animals. We identify the contradictory ways in which these wild urban animals are represented as both 'troublemakers' and 'cute' within children's literature, but also within the adult discourses that circulate about living with possums and raccoons in cities and discuss the effects of all of this upon children's real-life encounters with these animals. Unlike the shy, docile urban kangaroos and deer, urban raccoons and possums are unruly, intrusive, and pesky. They cause trouble. So, in this chapter, we borrow from Haraway's (2008) notion of the discomforting multispecies 'contact zone' and adopt her ethics of 'staying with the trouble' (Haraway 2016) as a way of cohabiting in unruly, discomforting, and unpredictable common worlds.

In the sixth and final substantial chapter, 'Indigenous child-dog relations: a recuperative ethics of kinship obligation', we reflect upon the responsibilities and obligations inherent in indigenous children's kinship relations with dogs. We draw the material for this chapter from indigenous accounts of Inuit children's relations with qimmiit (sled dogs) in the Canadian Artic and Arrernte and Warlpiri children's relations with camp dogs in the Central Australian desert. We are particularly interested in these accounts because, in pedagogical, cosmoecological, and ethical terms, these indigenous children's relations with dogs exceed limited western notions of both kinship and of ethics. We explore the ways in which these indigenous notions of child-dog kinship and ethics resonate with Donna Haraway's (2017) notion of 'oddkin', Thom van Dooren's (2007, 2014) discussions of multispecies intergenerational inheritances, and Vinciane Despret and Michel Meuret's (2016) notion of an 'ecology of obligations'. Drawing links between all of these trans-species reconceptualisations of kinship and ethics, we discuss how a more-than-human recuperative ethics of kinship obligation might offer hope within a diversity of damaged common world settings.

In the concluding chapter, 'Relational ethics for entangled lives', we return to Latour's (2004, p. 98) challenge of how to 'connect the question of the common world to the question of the common good' by summarising the range of ethical possibilities we have entertained in the previous chapters. We note that we are not trying to prescribe a one-size-fits-all universal ethical code. However, we are trying to promote an unfolding and relational common worlds ethics that is attuned to the profound global/local ecological challenges facing children and animals, and to the complex entanglements of their lives.

Notes

1 John Horton and Peter Kraftl (2006) were the first to challenge children's geographers to move beyond human-centric concerns.

2 Although the sub-discipline of children's geographies has been slow to engage with these 'turns', the work of John Horton and Peter Kraftl is a notable exception. See Horton and Kraftl (2006) and Kraftl (2013). Within the discipline of human geography, Kay

Anderson (1995), Sarah Whatmore (2002), and Jamie Lorimer (2010) have championed animal, hybrid, and more-than-human geographies, respectively.
3 See also Anna Tsing's (2015) similar methods of offering a 'rush of stories' from the 'unruly edges' of the 'capitalist ruins'.

References

Anderson, K., 1995. Culture and nature at the Adelaide Zoo: At the frontiers of 'human' geography. *Transactions of the Institute of British Geographers*, 20, 275–294.
Cole, M. and Stewart, K., 2014. *Our children and other animals: The cultural construction of human – Animal relations in childhood*. London: Routledge.
Despret, V. and Meuret, M., 2016. Cosmoecological sheep and the arts of living on a damaged planet. *Environmental Humanities*, 8(1), 24–36. doi:10.1215/22011919-3527704.
Haraway, D.J., 2004. Otherworldly conversations; terran topics; local terms. In: *The Haraway reader*. New York: Routledge, 125–150.
————., 2008. *When species meet*. Minneapolis: University of Minnesota Press.
————., 2016. *Staying with the trouble: Making kin in the Chthulucene*. Durham: Duke University Press.
————., 2017. *Making oddkin: Story telling for earthly survival*. Public Lecture, Yale University, October 23. Available from: www.youtube.com/watch?v=z-iEnSztKu8 [Accessed 9 February 2018].
Hird, M.J., 2010. Meeting with the microcosmos. *Environment and Planning D: Society and Space*, 28, 36–39.
————., 2013. Waste, landfills, and an environmental ethics of vulnerability. *Ethics and the Environment*, 18(1), 105–124.
Horton, J. and Kraftl, P., 2006. What else? Some more ways of thinking and doing 'children's geographies'. *Children's Geographies*, 4(1), 69–95.
Kahn, P.H. and Kellert, S.R., eds., 2002. *Children and nature: Psychological, sociocultural, and evolutionary investigations*. Cambridge, MA: MIT Press.
Kraftl, P., 2013. Beyond 'voice', beyond 'agency', beyond 'politics'? Hybrid childhoods and some critical reflections on children's emotional geographies. *Emotion, Space, and Society*, 9, 13–23.
Latour, B., 2004. *The politics of nature: How to bring the sciences into democracy*. C. Porter, trans. Cambridge, MA: Harvard University Press.
————., 2011. *Waiting for Gaia: Composing the common world through arts and politics*. Lecture, French Institute, London, November 2011. Available from: www.bruno-latour.fr/sites/default/files/124-GAIA-LONDON-SPEAP_0.pdf [Accessed 9 February 2018].
————., 2014. Another way to compose the world. *Journal of Ethnographic Theory*, 4(1), 301–307.
Le Guin, U.K., 1989. *Dancing at the edge of the world: Thoughts on words, women, places*. New York: Grove Press.
————., 1996. The carrier bag theory of fiction. In: C. Glotfelty and H. Fromm, eds. *The ecocriticism reader: Landmarks in literacy ecology*. Athens: University of Georgia Press, 149–154.
Lorimer, J., 2010. Moving image methodologies for more-than-human geographies. *Cultural Geographies*, 17(2), 237–258.
————., 2015. *Wildlife in the Anthropocene: Conservation after nature*. Minneapolis: University of Minnesota Press.

McCardle, P., McCune, S., Griffin, J.A. and Maholmes, V., eds., 2011. *How animals affect us: Examining the influence of human – Animal interaction on child development and human health*. Washington, DC: American Psychological Association.

Melson, G., 2005. *Why the wild things are: Animals in the lives of children*. Cambridge, MA: Harvard University Press.

Myers, G., 2007. *The significance of children and animals: Social development and our connections to other species*. 2nd ed. West Lafayette: Purdue University Press.

Puig de la Bellacasa, M., 2010. Ethical doings in naturecultures. *Ethics, Place, and Environment*, 13(2), 24–36.

Rose, D.B., 2004. *Reports from a wild country: Ethics for decolonisation*. Sydney: UNSW Press.

―――., 2011. *Wild dog dreaming: Love and extinction*. Charlottesville: University of Virginia Press.

―――., 2012. Multispecies knots of ethical time. *Environmental Philosophy*, 9(1), 127–140.

Ross, S.B., 2011. *The extraordinary spirit of green chimneys: Connecting children and animals to create hope*. West Lafayette: Purdue University Press.

Tsing, A.L., 2015. *The mushroom at the end of the world: On the possibility of life in the capitalist ruins*. Princeton: Princeton University Press.

van Dooren, T., 2007. Terminated seed: Death, proprietary kinship, and the production of (bio)wealth. *Science as Culture*, 16(1), 71–93.

―――., 2014. *Flight ways: Life and loss at the edge of extinction*. New York: Columbia University Press.

van Dooren, T. and Rose, D.B., 2012. Storied places in a multispecies city. *Humanimalia: A Journal of Human/Animal Interface Studies*, 3(2), 1–27.

Whatmore, S., 2002. *Hybrid geographies*. Thousand Oaks: SAGE.

Wilson, E.O., 1984. *Biophilia*. Cambridge, MA: Harvard University Press.

1   The common worlds of children and animals

> If you want to clear a room of derrideans, mention Beatrix Potter without sneering. . . . In literature as in real life, women, children and animals are the obscure matter upon which Civilization erects itself, phallologically. . . . If Man vs Nature is the name of the game, no wonder the team players kick out all these non-men who won't learn the rules and run around the cricket pitch squeaking and barking and chirping!
>
> (Le Guin 1990 cited in Emel and Wolch 1998, pp. 16–17)

Like Ursula Le Guin (1990), we take defiant pride in paying serious attention to the common worlds of children and animals that are shunned by the 'team players' and in writing about the 'obscure matter' of these worlds. We enjoy hanging out with children and animals in our empirical research and reading children's animal texts, precisely because the material/semiotic menageries that constitute children's and animals' mixed-up common worlds confound the rules of the main game – the hermetically sealed dualist formula of 'Man vs Nature'. We are acutely aware that children and animals, both fictional and real, are usually relegated to the 'nature' side of the divide, where they can be easily trivialised and/or dismissed by the adult-centric phallologics of 'serious' theory. Moreover, positioned on the side of 'Nature', children and animals are prone to be romanticised and idealised by those who look for refuge or escape from the corrupting influences of 'Man'.

The romantic coupling of children and animals under the rubric of 'pure nature' is easy to find within Jean-Jacques Rousseau's highly influential 18th-century philosophies (Rousseau 2003 [1762]). In Europe, it found expression in the 18th- and 19th-century romantic art and literature movement that revived the pastoral idyll and which typically depicted children and animals frolicking together in utopian scenes of rural tranquillity and harmony (Holloway and Valentine 2000). Across the Atlantic, in the 'New World' colonies, Rousseau's philosophies influenced the 19th-century North American transcendentalist movement that reified the 'wilderness' ideal (instead of the gentler rural idyll) and aligned children's natural instincts with those of wild animals (Thoreau 2009 [1862]). The transcendentalists looked to childhood as a way of reconnecting with the 'animal within' – as the link to reviving our primaeval connection with nature (Aitken 2001, p. 33).[1]

Within and across fields of childhood studies, children's geographies, and environmental education, these romantic traditions still hold traction. There is a perceptible trend to sanctify and idealise the relationship between children and animals as one of harmony, purity, and innocence. This has been achieved by casting children as instinctual, prerational, asocial, apolitical, and uncorrupted beings, and therefore aptly positioned alongside animals on the 'nature' side of the nature/culture divide (Taylor 2017).

Our vision for the common worlds of children and animals has nothing to do with harmony, purity, and innocence – either in the form of bucolic pastoral idylls or essentialised wild nature fantasies. It is much more pragmatic and down to earth. We see the entangled common worlds that children and animals inherit and co-inhabit as messy and mixed up rather than pure, as damaged rather than utopian, and as prosaic rather than sanctified. However, we also see the warts-and-all everyday common worlds of children and animals as lively, dynamic, and brimming with potential.

As we noted in the introduction, 'common worlds' is a term we borrow from Bruno Latour. Latour (2004, 2011, 2014) insists that the common world is neither predetermined nor fixed, but in a continuous state of composition, of commoning. Moreover, he stresses that humans are not the sole composers or caretakers of the commons. His active and more-than-human collective notion of common worlds is akin to Donna Haraway's (2008, 2016) generative, agentic, and interspecies notion of 'worldings' – or the co-making of worlds. Following Latour and Haraway, we like to think of the common worlds of children and animals as the ongoing, interactive, multispecies process of becoming worldly together. Herein lies their potential.

We begin this chapter by discussing the collective process through which we have assembled our *thinking through the common worlds of children and animals*. We take a lead from Donna Haraway here. Paying due regard to what she has learnt from Marilyn Strathern, Haraway (2016, p. 12) reminds us that 'it matters what ideas we use to think other ideas (with)'. She likens the collective process of sharing useful key concepts and ideas as a form of 'sympoiesis', or 'making-with', and refers to the ideas of others as much-needed 'gifts' (Haraway 2016, p. 5).

The key ideas we use to think with (and the list is by no means finite) come from feminist science studies, more-than-human and animal geographies, and multispecies studies within the environmental humanities. Ideas such as these have emerged from the academy in an effort to redress the anthropocentric limits of Euro-western humanist knowledge traditions and are associated with a series of decentring conceptual shifts variously referred to as the 'posthuman turn', the 'animal turn', the 'ontological turn', and the 'affect turn'. However, we note that not all of the ideas that help us to think about the common worlds of children and animals come from the academy. Our on-the-ground research with children and animals has also shaped our thinking, as have our relationships with indigenous peoples in Canada and Australia, whose knowledge comes directly from their 'land' or 'country' (Turner 2010; TallBear 2017).[2]

After paying our dues for our common world thinking, we move on to outline the big-picture global issues and concerns that frame the situated studies of the following chapters. These discussions articulate how and why macro geo-political, ecological, cultural, and ethical matters are relevant to the micro everyday lives of children and animals. We start by considering some compelling planetary *common world ecological inheritances* that cannot be ignored and which are the reason for our claim that the common worlds of children and animals are neither pure and innocent nor utopian. We consider the cascading sets of ecological legacies associated with the Anthropocene, the proposed name for a new geological era of 'mankind' (Crutzen 2002) precipitated by the fossil-fuel-burning enterprises of capitalist modernity. Noteworthy amongst the indicators of the Anthropocene are human-induced climate change and the unprecedented acceleration of human-precipitated species extinctions, both of which have direct implications for the common worlds (and futures) of children and animals. In considering the ecological inheritance of Canadian and Australian children and animals in our local common world studies, we reflect upon the legacies of settler colonialism and the enmeshment of European empire building with early modernity's drive towards 'expansion', 'progress', and 'development'.

In the *Travelling Animal Stories* section, we shift our focus to the cultural 'legacies' that mediate child–animal common world relations. We discuss the ways in which animal stories travel between the global and the local. More specifically, we look at the ways in which classic anglophone children's animal stories are implicated within imperial circuits and networks of exchange. We note that these geo-culturally specific stories, and their subtexts, differentially position certain children in relation to certain animals and thereby to the 'real' animals they live with.

In the final section – 'Common World Lives' – we reflect back on the specificities of child–animal relations on the everyday grounds of our respective Australian and Canadian local common worlds. We briefly explain how we conduct our ethnographic studies of child-wildlife encounters and interactions in urban settings and how we approach them as simultaneous shaped by and potentially reshaping uneasy settler multispecies relations in the contact zone (Haraway 2008).

Thinking through the common worlds of children and animals

With feminist science studies scholars

The process of thinking through the common worlds of children and animals has been collective and incremental. All of the assorted ideas we have borrowed and incorporated are part of our malleable tool bag. However, it is particularly packed with the wild and generous feminist philosophies of the biologist turned science studies scholar Donna Haraway. We are truly indebted to Haraway for our own thinking practices. We have been devouring her pioneering boundary-crossing ideas and utilising her quirky concepts for many years, and they continue to turn

our thinking inside out. It was in Haraway's (2003) writings that we first encoun-
tered a way of performatively thinking and writing nature and culture together,
instead of endlessly critiquing (and thus inadvertently reiterating) the nature/
culture divide. Her composite concept 'naturecultures' not only subverts and con-
founds this divide, but it simultaneously performs the entanglement of worlds – in
which everything deemed 'cultural' is simultaneously 'natural' and everything
deemed 'natural' is simultaneously 'cultural'.

We also appreciate how 'naturecultures' disrupts the humanist binaries that
define who 'we' are. It recomposes humans and the rest, minds and bodies, subjects
and objects, semiotic and material worlds. 'Naturecultures' not only underpins
the ways we think about common worlds, but also how we go about recomposing these worlds on the grounds of our own research with children and animals.
Haraway also helps us to understand that the ideas we think with, the words we
use, and the meanings we give to things have material consequences. This is why
ideas matter and why careful choice is so important. As she puts it, 'The word is
made flesh in mortal naturecultures' (Haraway 2003, p. 100).

As we have just indicated, 'entanglements' is another hallmark Haraway concept that accompanies naturecultures and which we, along with so many others,
can no longer think without. It is liberally scattered throughout this book. Haraway uses it, not only as an antonym to separations but also as a kind of ethical
mantra. Moreover, her constant emphasis on the 'com' or the 'with' (as in *companion species*', '*com*posing', and '*becoming with*') underscores her instance that
'no species acts alone' (Haraway 2015, p. 159) – we always *com*pose or make
worlds *together*. Her ideas and concepts are particularly appealing to think with
because they are both profoundly helpful and great fun. They carry considerable
weight in a joyful and light-hearted kind of way – the stamp of her celebratory
and irreverent feminist style. Her method of 'bag lady story-telling' (Haraway
2004) that we discuss in the introduction is a case in point. There are simply far
too many indispensable Haraway ideas to mention here, but they exert a powerful
presence throughout this book.

Isabelle Stengers is another feminist science studies scholar whose ideas and
ways of thinking stretch our own in ways that afford a fair bit of pain but also
much pleasure. In particular, our thinking and our research practice have been
driven by a 'Cosmopolitics' lecture Stengers gave in Canada in 2012, in which she
floated the simultaneously daunting and thrilling possibilities of 'collective thinking in the presence of [nonhuman] others' in order to produce 'common accounts'
of the world. In Stengers' (2010, 2018) terms, to think collectively requires us to
risk interrupting our entrenched (humanist) habits of thought and to slow down
in order to notice what else is going on and what is at stake for all who belong
in this world. For those of us thoroughly schooled in the binaries of the Euro-
western episteme, there is a huge amount of unlearning to do before we can make
the shift to collective thinking and common accounting – not the least, learning
how to relinquish the subject-object 'I know about . . . presumption of human
exceptionalism.

With children

Undertaking research with preschool-aged children guides us in this challenging task, as collective thinking is not necessarily so hard for them. Many seem unconstrained by the subject-object categorical boundaries – for instance, when they enthusiastically throw themselves into 'becoming with' other animals in their play. Before they embark on formal schooling, young children appear less enculturated into the practice of separating off from the world in order to know *about* it from a distance. This shows in moments when the small details and goings-on of life around them draw them in and absorb them in unselfconscious, affective, and embodied ways. Witnessing such small but significant events during our ethnographic research, in tandem with the ideas we take from Haraway and Stengers, helps to materialise our understanding of the *how* of 'becoming with' and of 'collective thinking in the presence of others'.[3]

With (more-than) human and animal geographers

Like science and technology studies, the broader discipline of geography straddles the 'social' and the 'natural' sciences. This positioning predisposes many human geographers to think across the nature/culture divide and to reconceptualise geography as a more-than-human affair (Whatmore 2006; Greenhough 2014). The impetus for 'rethinking the human of human geography' (Whatmore 1999) has come from a few different directions, and we briefly trace the threads that have most influenced our thinking.

From the early days of the 'animal turn', human geographers undertaking 'animal geographies' have been highlighting the material and discursive ways in which human and (nonhuman) animal lives are entangled 'in the nature-culture borderlands' (Emel and Wolch 1998), exploring the implications of animal agencies (Philo and Wilbert 2000), and considering the roles that animals play in the humanist boundary-making project (Philo and Wilbert 2000). Kay Anderson's (1995) study of the Adelaide Zoo is a standout in this latter regard. In this pioneering article, she reflects upon the ways in which the zoo enshrines and popularises the Cartesian boundaries of humanity and animality, and seems to epitomise the 'human capacity for order and control' of the exteriorised animal world (Anderson 1995, p. 292). In the settler colonial context of the Adelaide Zoo, it flexes 'the muscle of colonial mastery over nature' (Anderson 1995, p. 292). For Anderson, the zoo is implicated both materially and symbolically 'in the imperial network of animal trading' (Anderson 1995, p. 281), as the imperial taxonomies that further distinguish hierarchical categories of exotic animals have been used to fortify the racialisation of the human species. Her insights are particularly pertinent for us as we research and write within our respective settler-colonised lands. Anderson's thinking has sharpened our appreciation of the ways in which animals of empire have been used in children's fiction, a consideration we return to later in this chapter.

More recently, animal geographies have shifted attention to the ecological significance of human-animal relations and to questions of method. Jamie Lorimer's

(2015) call for a reconceptualisation of wildlife conservation in a 'post-natural' world has informed our ethnographic studies of urban child-wildlife relations. In addition, his attunement to the methodological and ethical implications of affect within human-animal relations (Lorimer 2014) has helped us to sharpen our ethnographic focus upon the ways in which children and animals affect each other and the ethical possibilities that flow from their multisensory encounters.

Over the last two decades, (more-than) human geographers have made incremental shifts in thinking about the relationship between the matter of the world (the geos) and life within it (the bios). This can be seen from the early calls to incorporate nature back into our conceptualisation of the social via the composite notion of 'socionature' (Braun and Castree 1998; Castree and Braun 2001) and to practise 'hybrid geographies' (Whatmore 2002), through to Nigel Clark and Kathryn Yusoff's (2017) current challenges to the core distinction between life (bios) and non-life (geos). Under the rubric of 'geosocial formations', they are playing with the possibility of *in*human ontologies by exploring the geomorphic, rather than other active life forces (or vitalism), as providing the conditions of possibility for human existence.

While this may pose a radical challenge to the dualisms of the Euro-western episteme, we note that within the indivisible indigenous knowledge traditions of North America and Australia (at least), there is no prerequisite sense of 'inert' to support a life versus non-life dichotomy. Land is life giving, and land and life are inseparable. In lieu of dualist epistemologies, there are only 'geontologies' – ontologies that emerge from conjoint geological and biological substance, composition, and obligation (Povinelli 2016).

Within the academy, even a basic recognition that the 'geos' matter in tandem with the 'bios', and that we need to think them both together, has the potential to decentre humanism's autonomous individual. While conducting our multispecies ethnographies on the local grounds of two very geographically distinctive British settler-colonised lands – in the cold and wet temperate coniferous urban forests of coastal British Columbia and the hot and dry urban grassy woodlands of the Australia Capital Territory – we have been constantly reminded of Chris Philo's (1992) affirmation that geographies displace histories, simply because things turn out differently in different places. For this reason, we have deliberately structured the following chapters around paired accounts of child–animal relations that we have witnessed in our respective research sites. This allows us to explore exactly *how* place matters in the unfolding of child–animal relations and to showcase how and why the 'where' of settler colonisation makes a difference.

The place theorist with the most formative influence on our thinking is Doreen Massey. From her earlier reconceptualisations of local place as an 'extraverted' constellation of globally networked 'power geometries' (1993), to her later thinking about place as a mutable, power-laden event of human and nonhuman 'throwntogetherness' (Massey 2005), Massey's ideas have shaped the ways in which we think about the constitution of common worlds and their uneven grounds. Her notion of the throwntogetherness of place, in particular, resonates with and informs our engagements with local places in settler-colonised lands. It also supports our practice of tracing the constellating material/semiotic and geo/ historical trajectories that throw children and animals together on local common world grounds.

With multispecies environmental humanities scholars

The most recent body of thinking to influence ours comes from the environmental humanities, a field that has been gaining momentum in response to growing awareness about anthropogenic (literally 'humankind') destabilisation of the earth's ecological systems. The planetary scale of these catastrophic changes and the crises that they are precipitating has prompted calls to 'unsettle' the human-centric and human exceptionalist premises of the humanities and to seriously reconsider what it means to be human by 'thinking through the environment' (Rose *et al.* 2012). As multispecies scholars within the environmental humanities point out, thinking through the environment includes repositioning ourselves as just one species amongst many that make and remake worlds (van Dooren and Rose 2012; Kirksey 2014; van Dooren 2014). This is a move that we are making as we approach the entangled relations between children and animals in common worlds as mutually affecting and co-constitutive.

Proponents of multispecies studies are also championing new modes of inquiry that refocus on what is already going on in the world beyond the human and upon the interdependent relations between diverse lifeforms. In their editorial introduction to a special 'multispecies' issue of *Environmental Humanities*, Thom van Dooren *et al.* (2016, p. 1) propose that 'cultivating arts of attentiveness' is a prerequisite for such research. They argue that exclusively human stories cannot take account of the complex ways in which all forms of life 'bring one another into being' or 'become in consequential relationship with others' (van Dooren *et al.* 2016, p. 3). They urge researchers to cultivate new, collaborative, and above all 'immersive' forms of inquiry that are crafted around paying close and careful attention to 'knowing and being' with nonhuman others, and to recognising 'what matters to them' (van Dooren *et al.* 2016, p. 6).

Our multispecies ethnographic inquiries, in particular our 'walking with wildlife' methods (Taylor and Rooney 2017), are attuned to cultivating such attentiveness with children. We are particularly taken by what Anna Tsing (2015, pp. 17–26) refers to as the 'arts of noticing' unassuming but resilient human and nonhuman 'polyphonic assemblages' that survive and quietly thrive in the 'unruly edges' of the 'capitalist ruins'. It is within these marginal, trashed, and abandoned spaces that Tsing looks for small but hopeful signs of multispecies recuperation from the ecological destruction wrought by the human progress projects of colonialism and capitalism. Working with children and animals, we identify with her emphasis on deliberately noticing what is going on in the sidelined worlds of overlooked minor players, for we too believe in the transformative potential of mutually enriching human-nonhuman encounters and relations, no matter how small and seemingly insignificant. Tsing's (2015, p. 37) method of listening to and telling 'a rush' of unruly stories resonates with us. She offers her rush of stories as a way noticing and noting how the 'indeterminate encounters' of diverse assemblages produce life. In a similar vein, the following chapters are full of unruly stories and vignettes that showcase how indeterminate encounters between children and animals can throw light on the possibilities of living well together.

Common world ecological inheritances

Inheriting the legacies of the Anthropocene[4]

By now it should be crystal clear that we do not see the common worlds that children and animals inhabit as Disneyland-like utopian worlds where pure and innocent small creatures frolic away from the troubles of the 'real' world. The entangled lives that we refer to are prosaic, messy, and imperfect ones, as are the sites of encounter. There are global/local trajectories that precede all encounters, along with their legacies. There is no outside. The lives and futures of the children and animals that inherit these legacies are already implicated in them. In terms of futurity, we believe that the most pressing and challenging legacies that children and animals inherit are ecological – and they are of a planetary scale, although experienced in very specific ways in the local. Beyond our particular focus upon children and animals, these same planetary ecological inheritances are increasingly being framed in terms of the Anthropocene.

Ever since its popularisation, at the turn of the 21st century, by the Nobel prize-winning scientist Paul Crutzen, the term Anthropocene has become a pivotal concept spawning widespread debate within and beyond the physical sciences. Crutzen (2002) chose this term, which literally translates as the 'Age of Man', to encapsulate the proposition that we are entering into a new geological epoch in which human activities have fundamentally and permanently changed the earth's geo-biospheric systems. Advocating for an official recognition that we are transitioning from the Holocene into the Anthropocene, earth scientists argue that humans first started to become 'a global geophysical force' during the industrial revolution and that this anthropogenic force moved into what they call 'The Great Acceleration' at the end of World War II (Steffen *et al.* 2007). The case they present for the name change is a series of calculations which point to the accelerating impacts of human activities upon the earth's interconnected systems, including the acidification of oceans; the depletion of the ozone layer; fundamental changes to the earth's carbon, phosphorous, and nitrogen cycles; global warming and climate change; and the rapid loss of biodiversity. It is the extent of biodiversity loss, evidenced by an unprecedented and exponentially increasing rate of anthropogenic species extinctions, and also referred to as the 'sixth mass extinction event' (Ceballos *et al.* 2015), which provides the most compelling evidence of anthropogenic biospheric change.

The cumulative body of scientific evidence that human actions have caused fundamental and increasingly irreversible earth systems changes raises the question of the viability of life on earth as we know it, including the probable extinction of our own (human) species along with so many others. It is this radically altered and precarious world and its concomitant uncertain futures that we now inherit and bequeath to future generations. This realisation compels us to consider our ethical responsibilities, as scholars working with children and animals, to tackle the pressing, interrelated questions of interspecies and intergenerational justice in this age of intensifying anthropogenic damage. We see this connection between intergenerational and interspecies justice as an inexorable one. All of the thinking we have acknowledged earlier as informing our own, and which leads us to situate multispecies relations with the complex mix of human and nonhuman, material and semiotic assemblages, or common worlds, reinforces our sense that humans and other species share, not only entangled, cascading, and enmeshed pasts but also the present and future Anthropocene legacy.

The additional understanding that anthropogenic environmental damage has been wrought by *specific* human projects, such as colonisation, industrialisation, and modernity, all of which fall under the irrevocably tainted banner of human 'progress and development', intensifies calls within the social sciences and humanities for a complete paradigm shift in thinking about what it means to be human and our place and agency in the world. We read the Anthropocene, not only as a wake-up call to the solipsism of the Euro-western progress and development project but also as a validation of entanglement and interdeterminacy. The scientific evidence confirms that we live in a world in which 'natural and human forces' are so complexly intertwined that 'the fate of one determines the fate of the other' (Zalasiewicz *et al.* 2010, p. 2231). This makes it very apparent that the bifurcations of western epistemologies that relegate 'nature' in a separate realm to 'culture' are now completely untenable. It is no longer possible to separate the human from the so-called natural world and no longer plausible to deny that our fate, as a species, is bound up with the fates of other species (Rose *et al.* 2012).

The proposed naming of the Anthropocene, and the debates that it raises, are fraught with paradoxes. As an undeniably powerful, creative, and destructive species, humans are not only responsible for but also mortally vulnerable to the life-threatening systemic changes that threaten our future existence, as well as that of countless other species. On the one hand, the omnipotent belief that we are an exceptional species appears to be reconfirmed by the declaration of the Anthropocene. However, as the life-threatening implications of the Anthropocene make clear, this same abiding faith in human exceptionalism is also our undoing. It traps us within the delusional beliefs that we can endlessly intervene to 'improve on nature', that we will always find a new technological 'fix' for the damaging fallout from our previous fixes, and that we can continue to deplete the earth's resources with impunity. It is the fantasy of human exceptionalism that predisposes us to disavow our mortal entanglement in the same earth systems we so radically disturb. To put it another way, it is a solipsistic belief in human exceptionalism and our hyperseparation from the rest of the world that derailed us in the first place (Plumwood 1993) and has subsequently produced the planetary upheaval that is now being referred to as the Anthropocene.

We join the growing numbers of feminists who critique the triumphalism and paradoxes of the Anthropocene as a naming event and call for a well-considered response (for instance, Gibson *et al*. 2015; Haraway *et al*. 2015; Tsing 2015; Tsing *et al*. 2015; Haraway 2016; Stengers 2016).

We take it as a wake-up call and a transformational opportunity. We have no doubt that it would be counterproductive if the Anthropocene simply became an excuse to ramp up human domination and control, to scramble for yet another grandiose human techno-fix, and to construct yet another heroic salvation narrative of (good, human) environmental stewards to the rescue (Haraway 2015). At the same time, business as usual is no longer enough. Something needs to change. Along with other feminist scholars, we seize the naming of the Anthropocene as a critical moment for new interventions and a shift in research focus.

Our first intervention involves situating ourselves firmly within ecological systems (Gibson *et al*. 2015) and keeping ourselves there. To state that humans are part of the ecology is to state the obvious – in fact, it is a fundamental premise of the Anthropocene declaration. However, subsequent moves to quickly find new human 'solutions' to the Anthropocene crises slide straight back into separating ourselves off again. They tap directly back into the kind of supremacist and dualistic thinking that got us into this mess in the first place. To resist this slide back into separation, our common worlds research remains resolutely focused upon the multifarious ways in which children's everyday lives, fates, and futures are inextricably entangled with the lives of other animals.

Closely related to this, our second common worlds intervention is to reconceptualise belonging and agency. For us, the declaration of the Anthropocene reconfirms the dangers of treating the world as our (human) stage and seeing ourselves as the only actors on it. Central to our reconceptualisation of the common worlds is that they belong to all, that we (humans) are just one set of players amongst many, and that all forms of agency are collective and relational. From this position, we deliberately shift away from the kind of research that speaks 'on behalf of nature' and that seeks to foster individual agentic children to become environmental stewards through developing caring relationships with animals. Instead, we focus upon identifying the ethical possibilities inherent in the messy and fraught child–animal encounters, interactions, and relations that are already taking place in local common worlds in the face of the precarious global ecological futures that we all inherit and face together.

The third intervention responds to the limits of only considering human lives (or occasionally also the lives of other species admired for their 'intelligence') as worthy of ethical considerations and to the folly of denying the ethical implications of our entanglement with all other lifeforms. By 're-situating nonhumans within ethical terms' (Gibson *et al*. 2015), our research considers both the intergenerational and interspecies ethical implications of anthropogenic ecological legacies. To this end, we seek to identify and enact an inclusive common world ethics that does not begin and end with its human constituents.

Inheriting the legacies of settler colonialism

In our Australian and Canadian research contexts, as in so many other parts of the world, the geo-biospheric changes associated with the Anthropocene are inseparable from the ecological legacies of colonisation. As a geo-historical phenomenon, European colonisation of indigenous peoples' lands was not only justified as part of the inevitable 'great march of civilisation', but it was also a harbinger of the widespread land clearing and associated 'progress' and 'development' projects driven by industrial, consumptive, and extractive capitalism. The ecological legacies of all of this include loss of ancient indigenous land-care knowledges and practices, widespread deforestation, erosion, petro-chemical soil and water contamination, damage caused by mineral and fossil fuel mining, toxic waste, introduction of invasive species, altered river courses and micro-climates, and, of course, the precipitation of mass species extinctions – the strongest indicator of and contributor to the anthropogenic biospheric systems collapse that we are now witnessing.[5]

From our position, the ecological legacies of the Anthropocene that Canadian and Australian children and animals inherit can neither be conceived of nor responded to outside of the legacies of settler colonialism.[6] Moreover, these legacies are not only materially evident in blasted landscapes and irrevocably damaged biomes – they carry enduring symbolic weight as well. Even though the act of 'settling' lands seized from the original inhabitants is materially inscribed on the surface of the land, the legitimation of this settlement requires ongoing symbolic work. Lorenzo Veracini (2010) describes settler colonialism as a never-completed quest. Having physically and legally stolen/ secured the land, settler sovereignty can only be 'naturalised' through the process of establishing settler identification with and belonging to it. This means that settler myths and narratives of discovery, exploration, civilisation, and belonging are essential ongoing components of the (never complete) settler colonial project.

In settler societies, non-indigenous children are a target audience for these myths and narratives. Because they are perceived as the nation's future, the process of legitimising their naturalisation is particularly pertinent. Moreover, the characterisation of 'cute' native wild animals within children's settler narratives has long been a key strategy with the cultural politics of nation building. For well over a century, fictional native animal characters have been used to secure non-indigenous children's identification with and sense of belonging to the natural/ native environs of the colonised lands.[7] In recent years, as these children's narratives increasingly engage with ecological issues and the agendas of wildlife conservation and environmental protection movements, the emphasis is shifting from securing non-indigenous children's affections for native wildlife as a way of ensuring their sense of belonging to the nation, to interpolating them as the future environmental stewards and protectors of these native species.

Framed by the cultural politics of settler colonial societies, such non-indigenous child-native animal semiotic entanglements are a common feature of settler colonial inheritances. However, they are not all exactly the same. They are variously constituted by the geo/cultural/historical specificities of their local happenstances. In the chapters that follow, our task is to loosen and unravel some of the knotted settler colonial ecological legacies of indigenous dispossession, damaged 'settled' lands, native wild animals, and settler children within our situated Canadian and Australian studies.

Travelling animal stories

All child–animal relations are culturally mediated by animal stories, not just those in settler colonial societies (Cole and Stewart 2014). And clearly, not all of the animal characters are based upon real-life animals that children will encounter in their own locales. In the anglophone world of children's animal literature and popular culture, many animal representations are part of the continuing circuits and networks of the ex-British Empire. They travel back and forward between the imperial centre and the peripheries, and as they go they are diffracted through a series of geo-historically specific cultural lenses, with all their underpinning ideological subtexts. These complex semiotic exchanges are part of the time-space-mattering (Barad 2011) of diverse child–animal relations across far-flung parts of the world. In other words, travelling animal stories can be thought of as mobile cultural legacies which diffract and mediate common world child–animal relations on local grounds.

For instance, the large wild animals of India and Africa are most likely to first enter the imaginaries of Euro-western children through their exposure to classic colonial animal narratives – from Rudyard Kipling's late 19th-century *Jungle Book* stories, and their derivatives, through to animated films like Walt Disney Pictures' late 20th-century *Lion King*. Apart from occasional visits to the zoo, these children would have little opportunity to meet animals such as wolves, monkeys, tigers, and lions face-to-face. Chances are that the characterisations of these wild, exotic animals will be anthropomorphised; nevertheless, wild, exotic animals like these gain a sense of familiarity within these children's imaginaries. These wild colonial animal characters and tales tend to function as moral allegories. They also carry subtexts, if not of imperial conquest, at least of 'primitive' natural hierarchical order – of 'the law of the jungle'. These complexly layered messages about the wild natural order of the peripheries differentially position their child audiences, depending upon their cultural backgrounds and where they live.

Circulating in the opposite direction, from the imperial centre to the peripheries, famous characterisations of comparatively benign and often much smaller British 'wild' animals, such as Beatrix Potter's *Peter Rabbit* and ratty, toad, and badger from Kenneth Graham's *Wind in the Willows* have also travelled far and wide. They also become familiar fictional companions of children who may never have visited the British countryside, including children who live in the very different environments of the ex-British colonies. These animal characters are also anthropomorphised, full of moral tales and used as allegories for human behaviours, but they carry the subtexts of a 'cultured' and hence 'civilised' stratified society. Again, the messages these animal tales tell about the social order of the imperial centre differentially position their child readers depending upon their own locations.[8]

Winnie the Pooh is an exemplar of the multilayered material/semiotic circuitry of animals of empire within anglophone children's fiction. The character is based upon a real Canadian black bear, called Winnie, who ended up in the London Zoo. Winnie came to England from Canada in 1914 with Harry Colebourn, a member of the Canadian Army Veterinary Corp, en route to look after horses in the World War I Allied cavalry units in Europe, and was a mascot for his battalion. Harry named her Winnipeg Bear after the Canadian town that he lived in and left her in the London Zoo when he went to the battlefields of France in 1915. There she stayed for a further 19 years until her death. When visiting the London Zoo some years later, a boy called Christopher Robin Milne took a particular fancy to Winnie the Bear, thus inspiring his father, A.A. Milne, to write the classic stories about the adventures of a very cultured British bear and his boy companion.[9]

Although classically British, these Winnie the Pooh stories did not remain in the UK. They have been culturally diffracted, rearticulated, and widely redistributed in comic and animated film form by the Disney corporation in the USA and reproduced in soft toy, clothing, and game forms as part of the arsenal of children's commercial popular culture merchandise. The back story of how a real Canadian bear became the protagonist of a British children's fiction classic has also been made into a Canadian television film (Harrison 2004) as well as a children's picture book (Mattick and Blackall 2016). Diverse forms of the legendary Pooh Bear, both narrative and material, have been disseminated into all corners of the English-speaking world, including back to the settler colonial Canadian contexts from which the original Winnipeg Bear came.

A plethora of time-space-matterings (Barad 2011) has been produced during the global travels of this particular bear, each designed to secure children's affections for and identification with Winnie as a loveable animal companion. These stories of child-black bear companionship are a long way from the mortal relations between children and black bears in Canada and the USA, which are much more likely to pivot around fear, avoidance, and predation. And yet they are not entirely disconnected. For all of the globally circulating semiotic/material knots of children's and animals' lives, ecological and narrative, constellate in the thrown-together local moments of actual encounter (Massey 2005), regardless of the tensions and contradictions that they produce. In the chapters that follow, we trace these tensions – between the micro events of child–animal encounter in local common worlds and the macro geo-historical, cultural, and ecological processes and events that precede, intersect, and interpolate them.

Common world lives

It is a challenging task to trace exactly how global/local ecological and cultural articulations are played out within the everyday lives of children and animals – or what we are calling their common world lives. For us, this tracing process begins on the grounds of a series of university childcare centres in Canberra, Vancouver, and Victoria, British Columbia, where we have conducted our situated multispecies ethnographies[10] over the last five or six years. We simply observe, follow, and trace the relations that emerge between the children and animals that cohabit in these campus centres and in the nearby green spaces and neighbourhoods in which the children often walk. These are not staged relations, intentionally orchestrated as part of the children's education. In other words, they are not the same kind of subject-object relations that unfold when small caged animals deemed suitable for young children (like guinea pigs, rabbits, or hatching chicks) are deliberately brought in to childcare centres in order to provide the children with hands-on biological science education and to teach them how to care for domestic pets.

The child–animal relations we follow in our ethnographic studies are ones that evolve through happenstance encounter. Even though they are not predetermined relations, they nevertheless only occur because the various children and animals whom we observe interacting are already thrown together on these common grounds – for whatever reasons and via whatever trajectories. To put it another way, the children and animals concerned are local common world cohabitants who happen across each other in the course of their everyday lives and movements. We have witnessed a multitude of such happenstance child–animal encounters during years of ethnographic research. The ones that we recount in the following chapters – between children, worms, ants, deer, kangaroos, possums, and raccoons – are of particular interest to us because they are part of a pattern of repeated encounters within which both children and animals appear to be mutually affecting and affected. The animals and children act upon and move each other in discernible ways. Our focus is therefore upon everyday child–animal common world interactions that involve some degree of observable, ongoing co-transformation. These mutually transformative chance encounters – no matter how small and fleeting – offer us a glimpse into the kind of ordinary everyday interspecies relations that reshape worlds together in ways that exceed paramount human actions and interests.

Even though we follow the interspecies relations that emerge from interactions between children and animals who physically share common worlds, we remain cognisant of the material/semiotic mutability of these worlds and of their constitutive mobilities and displacements. We know that the inhabitants – human and other animals – are never just there. There are always traceable trajectories, involving much larger processes and forces, which bring them all together. The children, for instance, are from families who originated in far-flung parts of the world. They have all ended up attending these various university campus environs due to the geohistorical convergences of global processes, such as settler colonisation, diasporas, immigration, and the internationalisation of higher education. The animals have also converged on the various campuses through diverse routes and means. Only the ants and worms are in place. The deer, kangaroos, possums, and raccoons are amongst the increasing populations of urban wildlife displaced from their natural habitats by the ecological upheavals associated with colonialisation and the Anthropocene.[11]

Contact zones

The thrown-together common worlds of these children and urban wildlife are akin to what Haraway (2008, pp. 205–246) refers to as multispecies 'contact zones'.

Unlike the relatively predictable domestic living arrangements of children and 'pet' animals, which have a long-established cultural history, children and wildlife who are thrown together in the contact zone of local urban environments must negotiate the uncharted territories of newly unfolding relations, and in so doing, they affect each other in unpredictable ways.

Haraway borrows the term 'contact zones' from Mary Louise Pratt, who uses it to describe 'the interactive, improvisational dimensions of colonial encounters' (Pratt cited in Haraway 2008, p. 316). Like Pratt, Haraway insists that the interactions and encounters in the contact zone, although characterised by geo-historical power inequities and tensions, amount to more than simple relations of domination and subordination. Contact zones are uneasy but also productive spaces. The new relations they produce effectively reconstitute all of the subjects involved. Haraway also picks up on James Clifford's extended vision of these productive contact zone interactions as operating at 'intersecting regional, national and transnational levels' and as 'entering new relations through the historical process of displacement' (Clifford, cited in Haraway 2008, p. 217).

There is a strong sense that the cascading levels of displacement, from the global to local levels of common world contact zones, open new spaces of possibility. This is the sense in which Haraway takes the notion up and populates it with natureculture and multispecies entities and relations – as do we. We find the contact zone a very useful frame of reference for noticing what happens when children and wildlife meet in their immediate urban common worlds, and also a very useful reminder of the hopeful possibilities that reside in these potentially transformative exchanges.

Conclusion

This chapter has provided the back story to our thinking about the common worlds of children and animals, and how and why they matter – literally and to us. By acknowledging the key concepts and ideas that help us to circumvent the anthropocentric and bifurcating limits of humanist knowledge traditions, we have intended to render transparent the means by which we are seeking to better understand exactly how the lives of children and animals are mutually entangled. We have also sought to clarify why an appreciation of shared ecological legacies, coupled with an understanding of the inextricable entanglement of very different lives, are particularly pertinent in this time of growing awareness about anthropogenic damage to the earth's ecological systems and hence to the future of life on earth as we know it.

All of this requires sensitivity to global/local articulations and how they are formative of the common worlds of children and animals. In this chapter, we have sketched some of the patterns of global forces, circuits, and exchanges that constellate on local grounds and work themselves out in lives and relations of cohabiting children and animals. We hope that all of this provides clarifying context for the situated child–animal studies in the following chapters. It is these situated studies that bring substance to the big-picture approaches, patterns, and concerns we have outlined here.

Notes

1 For an extended genealogy of Rousseau's romantic 'nature's child' figure and the romantic legacies with early childhood and environmental education, see Taylor (2013b, pp. 3–16).
2 For a detailed account of how Affrica's relations with Arrernte and Luritja 'Yeperenye' (caterpillar) children of Mbantwe country (Alice Springs) have shaped my thinking, see Taylor (2013a).
3 Many written and visual accounts of such events can be found on the 'Walking with Wildlife in Wild Weather Times' research blog (Taylor and Rooney 2017).
4 For a similar but more detailed discussion of inheriting the legacies of the Anthropocene, see Taylor and Pacini-Ketchabaw (2015).
5 We note that Australia, as a continent, has suffered by far the highest level of mammal and plant species extinctions in the world in the last 200 years, or roughly since British colonisation (Woinarski *et al*. 2015).
6 For further discussion about the significance of settler colonial ecological inheritances for children and other animals, see Taylor *et al*. (2015).
7 The process of securing non-indigenous Australian children's identification with native animals and hence the Australian settler nation is more fully discussed in Taylor (2014).
8 Outside of the context of the circuitry of imperial animal stories, Leesa Fawcett (2002) writes about the bonds that children form with wild animals through narrative engagement.
9 There are many online historical accounts of this story, including one with original photographs by Christopher Klein (2016).
10 For an overview of the emergence of multispecies ethnographies, see Kirksey and Helmreich (2010). For a detailed discussion about the challenges of decentring the human in our own multispecies ethnographic studies, see Pacini-Ketchabaw *et al*. (2016).
11 Jamie Lorimer (2015) discusses the increasing phenomenon of urban wildlife as emblematic of life in the post-nature era of the Anthropocene. He suggests that wildlife is a way of rethinking nature beyond the urban/nature divide.

References

A bear named Winnie, 2004. TV film. Directed by J.K. Harrison. Toronto: CBD Home Video.

Aitken, S.C., 2001. *Geographies of young people: The morally contested spaces of identity*. London: Routledge.

Anderson, K., 1995. Culture and nature at the Adelaide Zoo: At the frontiers of 'human' geography. *Transactions of the Institute of British Geographers*, 20(3), 275–294.

Barad, K., 2011. Nature's queer performativity. *Qui Parle: Critical Humanities and Social Sciences*, 19(2), 121–158.

Braun, B. and Castree, N., eds., 1998. *Remaking reality: Nature at the millennium*. London: Routledge.

Castree, N. and Braun, B., eds., 2001. *Social nature: Theory, practice, politics*. Malden: Blackwell.

Ceballos, G., Ehrlich, P.R., Barnosky, A.D., Garcia, A., Pringle, R.M. and Palmer, T.M., 2015. Accelerating modern human-induced species losses: Entering the sixth mass extinction. *Science Advances*, 1(5), e1400253. doi:10.1126/sciadv.1400253.

Clark, N. and Yusoff, K., 2017. Geosocial formations and the Anthropocene. *Theory, Culture, and Society*, 34(2–3), 3–23.

Cole, M. and Stewart, K., 2014. *Children and other animals: The cultural construction of human – Animal relations in childhood*. London: Routledge.

Crutzen, P., 2002. Geology of mankind. *Nature*, 415, 23. doi:10.1038/415023a.

Emel, J. and Wolch, J., 1998. Witnessing the animal moment. *In*: J. Wolch and J. Emel, eds. *Animal geographies: Place, politics, and identity in the nature-culture borderlands*. London: Verso, 1–24.

Fawcett, L., 2002. Children's wild animal stories: Questioning interspecies bonds. *Canadian Journal of Environmental Education*, 7(2), 125–139.

Gibson, K., Rose, D.B. and Fincher, R., eds., 2015. *Manifesto for living in the Anthropocene*. Brooklyn: Punctum.

Greenhough, B., 2014. More-than-human geographies. *In*: R. Lee, N. Castree, R. Kitchin, V. Lawson, A. Paasi, C. Philo, S. Radcliffe, S.M. Roberts and C.W.J. Withers, eds. *The SAGE handbook of human geography*. Vol. 1. London: SAGE, 94–119.

Haraway, D., 2003. *The companion species manifesto: Dogs, people, and significant otherness*. Chicago: Prickly Paradigm Press.

———., 2004. Otherworldly conversations; terran topics; local terms. *In*: *The Haraway reader*. New York: Routledge, 125–150.

———., 2008. *When species meet*. Minneapolis: University of Minnesota Press.

———., 2015. Anthropocene, capitalocene, plantationocene, chthulocene: Making kin. *Environmental Humanities*, 6, 159–165.

———., 2016. *Staying with the trouble: Making kin in the Chthulucene*. Durham: Duke University Press.

Haraway, D., Ishikawa, N., Gilbert, S., Olwig, K.R., Tsing, A.L. and Bubandt, N., 2015. Anthropologists are talking – About the Anthropocene. *Ethnos*, 81(3), 535–564. doi:10.1080/00141844.2015.110583.

Holloway, S. and Valentine, G., 2000. Children's geographies and the new social studies of childhood. *In*: S. Holloway and G. Valentine, eds. *Children's geographies: Learning, living, learning*. London: Routledge, 1–28.

Kirksey, E., ed., 2014. *The multispecies salon*. Durham: Duke University Press.

Kirksey, E. and Helmreich, S., 2010. The emergence of multispecies ethnography. *Cultural Anthropology*, 25(4), 545–576.

Klein, C., 2016. The true story of the real-life Winnie-the-Pooh. *History Stories*, 13 October 2016. Available from: www.history.com/news/the-true-story-of-the-real-life-winnie-the-pooh [Accessed 25 February 2018].

Latour, B., 2004. *The politics of nature: How to bring the sciences into democracy*. Cambridge, MA: Harvard University Press.

———., 2011. *Waiting for Gaia: Composing the common world through arts and politics*. Lecture, French Institute, London, November 2011. Available from: www.bruno-latour.fr/sites/default/files/124-GAIA-LONDON-SPEAP_0.pdf [Accessed 12 February 2018].

———., 2014. Another way to compose the world. *Journal of Ethnographic Theory*, 4(1), 301–307.

Lorimer, J., 2014. On auks and awkwardness. *Environmental Humanities*, 4, 195–205. Available from: www.environmentandsociety.org/mml/lorimer-jamie-auks-and-awkwardness [Accessed 12 February 2018].

———., 2015. *Wildlife in the Anthropocene: Conservation after nature*. Minneapolis: University of Minnesota Press.

Massey, D., 1993. Power-geometry and a progressive sense of place. *In*: J. Bird, B. Curtis, T. Putnam, G. Robertson and L. Tickner, eds. *Mapping the futures: Local cultures, global change*. London: Routledge, 60–70.

———., 2005. *For space*. London: SAGE.

Mattick, L. and Blackall, S., 2016. *Finding Winnie: The story of the real bear who inspired Winnie-the-Pooh*. London: Orchard Books.

Pacini-Ketchabaw, V., Taylor, A. and Blaise, M., 2016. De-centring the human in multispecies ethnographies. *In*: C. Taylor and C. Hughes, eds. *Posthuman research practices in education*. Houndmills: Palgrave Macmillan, 149–167.

Philo, C., 1992. Foucault's geography. *Environment and Planning D: Society and Space*, 10, 137–161.

Philo, C. and Wilbert, C., eds., 2000. *Animal spaces, beastly places: New geographies of human – Animal relations*. London: Routledge.

Plumwood, V., 1993. *Feminism and the mastery of nature*. New York: Routledge.

Povinelli, E., 2016. *Geontologies: A requiem to late liberalism*. Durham: Duke University Press.

Rose, D.B., van Dooren, T., Chrulew, M., Cooke, S., Kearnes, M. and O'Gorman, E., 2012. Thinking through the environment: Unsettling the humanities. *Environmental Humanities*, 1, 1–5.

Rousseau, J.-J., 2003. *Emile, or treatise on education*. W.H. Payne, trans. New York: Prometheus Books (Original work published 1762).

Steffen, W., Crutzen, P. and McNeill, J.R., 2007. The Anthropocene: Are humans now overwhelming the great forces of nature? *Ambio: A Journal of the Human Environment*, 36(8), 614–621.

Stengers, I., 2010. *Cosmopolitics I*. Minneapolis: University of Minnesota Press.

———., 2012. *Cosmopolitics: Learning to think with science, people, and natures*. Public Lecture. Organised by Brian Noble and the Situating Science Knowledge Cluster, Halifax, Canada. Available from: www.youtube.com/watch?v=-ASGwo02rh8 [Accessed 12 February 2018].

———., 2016. *In catastrophic times: Resisting the coming barbarism*. London: Open University Press.

———., 2018. *Another science is possible: A manifesto for slow science*. S. Muecke, trans. Cambridge: Polity Press.

TallBear, K., 2017. Beyond the life/not life binary: A feminist-indigenous reading of cryopreservation, interspecies thinking, and the new materialisms. *In*: E. Kowal and J. Radin, eds. *Cryopolitics: Frozen life in a melting world*. Cambridge, MA: MIT Press.

Taylor, A., 2013a. Caterpillar childhoods: Engaging with the otherwise worlds of Central Australian Aboriginal children. *Global Studies of Childhood*, 3(4), 366–397.

———., 2013b. *Reconfiguring the natures of childhood*. London: Routledge.

———., 2014. Settler children, kangaroos, and the cultural politics of Australian national belonging. *Global Studies of Childhood*, 4(3), 169–182.

———., 2017. Romancing or reconfiguring nature? Towards common worlds pedagogies. *In*: K. Maloney, T. Gray and S. Truong, eds. *Reimagining sustainability education in precarious times*. Amsterdam: Springer, 61–75.

Taylor, A. and Pacini-Ketchabaw, V., 2015. Learning with children, ants, and worms in the Anthropocene: Towards a common world pedagogy of multispecies vulnerability. *Pedagogy, Culture, Society*, 23(4), 507–529.

Taylor, A., Pacini-Ketchabaw, V., Blaise, M. and de Finney, S., 2015. Inheriting ecological legacies of settler colonialism. *Environmental Humanities*, 7, 129–132.

Taylor, A. and Rooney, T., 2017. *Walking with wildlife in wild weather times*. A Common World Childhood Research Collective Blog. Available from: https://walkingwildlifewildweather.com [Accessed 12 February 2018].

Thoreau, H.D., 2009. *Walking*. The Thoreau Reader (Original work published 1862). Available from: https://thoreau.eserver.org [Accessed 25 February 2018].

Tsing, A.L., 2015. *The mushroom at the end of the world: On the possibility of life in capitalist ruins*. Princeton: Princeton University Press.

Tsing, A.L., Swanson, H., Gan, E. and Bubandt, N., eds., 2015. *Arts of living on a damaged planet: Ghosts and monsters of the Anthropocene*. Minneapolis: University of Minnesota Press.

Turner, M.K., 2010. *Iwenhe tyerrtye: What it means to be an Aboriginal person*. Alice Springs: IAD Press.

van Dooren, T., 2014. *Flight ways: Life and loss at the edge of extinction*. New York: Columbia University Press.

van Dooren, T., Kirksey, E. and Munster, U., 2016. Multispecies studies: Cultivating arts of attentiveness. *Environmental Humanities*, 8(1), 1–23.

van Dooren, T. and Rose, D.B., 2012. Storied places in a multispecies city. *Humanimalia: A Journal of Human/Animal Interface Studies*, 3(2), 1–27.

Veracini, L., 2010. *Settler colonialism: A theoretic overview*. London: Palgrave Macmillan.

Whatmore, S., 1999. Hybrid geographies: Rethinking the 'human' in human geography. *In*: D. Massey, J. Allen and P. Sarre, eds. *Human geography today*. Cambridge: Polity Press, 22–39.

———., 2002. *Hybrid geographies: Natures, cultures, spaces*. London: SAGE.

———., 2006. Materialist returns: Practising cultural geography in and for a more-than-human world. *Cultural Geographies*, 13(4), 600–609.

Woinarski, J.C.W., Burbidge, A.A. and Harrison, P.L., 2015. Ongoing unraveling of a continental fauna: Decline and extinction of Australian mammals since European settlement. *Proceedings of the National Academy of Sciences*, 112(15), 4531–4540. Available from: www.pnas.org/content/pnas/112/15/4531.full.pdf [Accessed 25 February 2018].

Zalasiewicz, J., Williams, M., Steffen, W. and Crutzen, P., 2010. The new world of the Anthropocene. *Environmental Science and Technology*, 44, 2228–2231.

2 Children, kangaroos, and deer
An ethics of multispecies conviviality

> The place of wildlife in the city opens our engagement with the urban in ethically compelling ways. The city is not so much an objective fact as it is a specific material mode of storying – a way of understanding relating and becoming. It is a story, told and enacted by many creatures.
>
> (van Dooren and Rose 2012, p. 18)

Urban stories of child, deer, and kangaroo cohabitation

How do children's lives intersect with those of wildlife in urban areas? How are children and wildlife co-implicated in inherited pasts and entangled presents and futures? What stories might bring these intersections and inheritances to life? How might these stories help us to better appreciate the complex ethical dimensions of human-wildlife cohabitation in urban places in ecologically challenging times? What ethical possibilities might reside in children's everyday encounters and relations with urban wildlife? What might we learn by telling stories of urban child-wildlife encounters and relations?

These questions can only be answered in very situated studies. In this chapter, we recount two geo-historically specific sets of child-wildlife stories of entangled inheritances and contemporary cohabitations. The first set of stories originates in the grassy suburbs of Canberra, Australia's young 'bush capital' city. Its focus is the interwoven trajectories of Australian children's and kangaroos' lives, and it includes narrative vignettes of urban child-kangaroo encounters drawn from Affrica's ethnographic research. The second set of stories comes from the forested suburbs of Victoria, the colonial capital of the province of British Columbia in western Canada. It similarly traces the multilayered ways in which the lives of Canadian children and deer intersect, and features vignettes of encounters between urban deer and children drawn from Veronica's ethnographic research. In both cases, these stories of urban multispecies entanglements are firmly situated within the fraught and very specific Australian and Canadian geo-histories of settler-animal relations and anthropogenic environmental damage.

As well as illustrating how child–animal relations unfold in distinctive ways in different places, these stories bear witness to a global trend. The combination of climate change, habitat destruction, and rapid urbanisation associated with colonisation and globalisation have resulted in the forced migration of many species, resulting in an increased concentration of wildlife cohabiting with humans in urban environments. Along with other scholars, we are responding to this trend by seeking a new ethics for multispecies living, an ethics that necessitates that we extend the notion of diverse belongings beyond our own species (van Dooren and Rose 2012; van Dooren 2014; Gibson *et al.* 2015; Lorimer 2015). To this end, the stories we present in this chapter not only showcase the specific ways in which children and wild animals share fraught colonial and ecological inheritances, they also portend hope. The vignettes of child-wildlife urban encounters are modest examples of the 'multispecies achievements' that are necessary for attaining 'conviviality', or 'the ethics of sharing places' (van Dooren and Rose 2012, pp. 3, 5). These ordinary, everyday, and seemingly insignificant interactions between children and kangaroos in Canberra, and children and deer in Victoria point towards the possibilities for an ethics of multispecies urban conviviality, which requires us to move beyond understanding the city as an exclusively human space (van Dooren and Rose 2012, p. 15).

Children and kangaroos: entangled inheritances

Since Australia's federation as a nation in 1901, the kangaroo has been a key symbol of national identity with special significance to young Australians. Along with the emu, it is one of the bearers on the Australian coat of arms. According to official government accounts, the kangaroo was chosen because it can only hop forward, thus signifying hope and optimism for this young (white) nation (Australian Government Department of Foreign Affairs and Trade 2013), but also firmly positioning Australian nationhood within the twin framings of progress and modernity. Tellingly, the emphasis on moving forward at this moment of transition from colonies to federated modern nation state also testifies to the repression or whitewashing of ancient indigenous cultural histories.

It was at least partly because of the whitewashing of Indigenous Australian human histories that the white-settler narratives of belonging so essential to the naturalisation process and the building of a national imaginary relied heavily upon the symbolic significance of native animals (Taylor 2013, 2014). As the emblematic national symbol, the kangaroo takes pride of place in Australian children's fiction and popular culture. Since federation, the figure of the kangaroo has been mobilised to secure settler children's affection for Australian bush creatures and thereby to consolidate their identification as naturalised Australian children.

Kangaroo characters in children's popular culture

Kangaroo characters abound in Australian children's popular culture, but two notable classics are Ethel Pedley's (1997 [1906]) turn of the 20th century *Dot and the Kangaroo* children's picture book, which was made into an animated film in the 1970s, and the 1960s children's television series *Skippy the Bush Kangaroo*, which was enormously popular at the time and has since gained cult status. In both instances, the narratives feature close relationships between a white Australian child (Dot and Sonny, respectively) and a canny kangaroo (Kangaroo and Skippy, respectively), and the kangaroo characters function as the children's guides and muses in the Australian bush. By highlighting the kangaroos' cleverness and the strong bonds between the children and the kangaroos, both these stories foster their young Australian audience's respect for and identification with the natural environment, and encourage conservation mindedness – a quality that is clearly associated with becoming good Australian citizens (Franklin 2006, p. 113).

In *Skippy*, the conservation message is communicated through a rollicking series of boy's-own and companion kangaroo bush adventures, but also by an insistence that, despite their close relationship, at the end of the day, Skippy the kangaroo is a wild animal who belongs in the bush. She never becomes Sonny's pet. In *Dot and the Kangaroo*, the relationship is definitely one of native (animal) muse and juvenile (human) apprentice. Kangaroo not only saves young Dot when she becomes lost in the bush, but she also helps her to see how Australian native animals have become the innocent victims of mindless white-settler cruelty. This narrative entreats Australian children to identify with the bush animals and to take the lead in protecting them. This message is made quite explicit in the foreword to the story, when Pedley (1997 [1906]) makes this direct plea: 'To the children of Australia in the hope of enlisting their sympathies for the many beautiful, amiable and frolicsome creatures of their fair land, whose extinction, through ruthless destruction, is being surely accomplished'.[1]

Fraught settler-kangaroo relations

Pedley's (1997 [1906]) reference to the possible extinction of Australian wildlife was not without context. In stark contrast to the affectionate ways in which the figure of the kangaroo was mobilised to promote Australian identity in the national imaginary, particularly in children's literature, flesh-and-blood kangaroos were widely regarded as vermin and subjected to widespread, systematic slaughter from the early colonial days through to the mid-20th century. In response to the perception that these fence-hopping, pasture-grazing kangaroos posed a major threat to the burgeoning sheep industry, marsupial destruction acts were passed in the new states. The scale and fervour of the ensuing state-sanctioned kangaroo hunting drives, mass shootings, and poisonings were akin to a state of war. In Queensland alone, there were reports that 27 million kangaroos were killed within 50 years (Simons 2013, p. 42). Australia wide, these 'kangaroo wars' resulted in the loss of many of the smaller and hence more vulnerable kangaroo species.

By the mid-20th century, a new breed of mostly urban-based conservationists was voicing concern about the rapid loss of Australian wildlife. As a result, kangaroos, along with other native animals, eventually gained the protected status that is still in place today. Wildlife protection legislation stipulates that kangaroos can only be killed under license if they pose a demonstrable threat to livestock farming or to other native species. As a consequence, kangaroo numbers have rebuilt and stabilised at around 57 million, mostly consisting of three large species, commonly known as the eastern greys, the western greys, and the reds (Simons 2013, p. 42). However, as the articulation of human and environmental forces comes increasingly into play, it is becoming clearer that this protected status is no fail-safe guarantee of kangaroo survival.

At the turn of the 21st century, a ten-year drought that gripped the tablelands of southeastern Australia was identified as the hottest and driest on record (SEARCH 2013), and offered the clearest evidence yet of the ways in which anthropogenic global warming and climate change might affect this bioregion. As this relatively temperate part of the continent heats up and dries out, there are signs that the eastern grey kangaroos that are indigenous to this part of the continent are beginning to struggle. Unlike the western greys and the reds that come from the more arid regions, the eastern greys have not evolved to live in such hot, dry climes. On top of this, the grasses that the settlers introduced for livestock 'pasture improvement', and which have now replaced most of the native grass species, are also very vulnerable to drought. Like the settlers' livestock, eastern grey kangaroos now depend on these introduced grasses for survival. It is lucky that they are such highly adaptable animals (Flannery 2008, pp. 24–25).

Kangaroos in Canberra

During the drought, unprecedented numbers of eastern greys moved from the surrounding over-cleared and drought-stressed sheep country of the southeastern tablelands into the urban precincts of Canberra. Finding sustenance in the bush capital's plentiful urban parks, grassy woodland nature reserves, curbside 'nature strips', and unfenced suburban front lawns, these immigrant urban kangaroos fared much better than their 'country cousins' during the drought. In fact, they adapted so well to life in the bush capital that they stayed on after the drought had broken. Canberra now has a permanent population of eastern grey kangaroos, and their numbers have exploded. According to local government sources, by 2010, Canberra had the highest density of kangaroos per hectare in the whole of Australia (Australian Capital Territory, Territory and Municipal Services 2010, p. 26), prompting the implementation of highly controversial annual kangaroo culls designed to manage kangaroo populations and reduce 'excessive grazing pressure' on the local remnant and fragile 'native grassy ecosystems' (Australian Capital Territory, Territory and Municipal Services 2010, p. 7).

Canberra's kangaroo culling controversy epitomises the fraught politics of multispecies belonging in urban environments. It points to the increasingly knotty complications that displaced wildlife bring to the question of belonging in the age of anthropogenic climate change and mass extinctions, and which intensify within the close living quarters of human purpose-built urban spaces (van Dooren and Rose 2012; Lorimer 2015). Before we move on to consider, as promised, how real-life encounters between children and kangaroos in Canberra might point to some hopeful possibilities for multispecies belonging, we want to recap a couple of points emerging from this discussion to date. First, we want to stress that even before the particular children and kangaroos in the forthcoming series of short stories met, their lives were unfolding in ways that were already shaped by the cultural and environmental legacies of Australian settler colonialism that they all inherited. Second, the potted geo-history we have just provided of fraught settler-kangaroo relations underscores that neither Canberra kangaroos nor children are *just self-evidently there*. They each have their own stories of 'coming to Canberra', which are contingent upon the (now undeniably entwined) human and inhuman forces that propel their constellating trajectories.

As Haraway (2008) puts it in *When Species Meet*, we are all engaged in a 'dance of relating' in which our 'sometimes-joined, sometime-separate heritages' come into play before and lateral to *any actual* encounter. In the vignettes that follow, we try and stay mindful of the children's and kangaroo's sometimes-separate, sometimes-shared inheritances, as well as open to the emerging co-shapings and becomings-with that are enabled through these face-to-face 'dances of relating'. In relation to the latter, this involves working with a new mode of attention – one that is far less cerebral and much more attuned to the embodied and sensorial ways in which children and kangaroos act upon, affect, and move each other.

Child-kangaroo encounters

The kangaroos are still clustered in the gully under the casurina saplings. That's where the spring grass is thickest and greenest. They haven't moved since we passed them on our way down to the bush cubby. We're on our way back now and pause to have another look at the mob. The children are always on the lookout for joeys.

Kangaroo spotting is a regular feature of our Wednesday afternoon bush walks. Roos are always somewhere to be seen in the forested area down behind the campus buildings. They're trapped on this ring-roaded campus, but at least they now have plenty to eat. The roos are definitely on the watch for us too. The standard routine is that they fixedly stare at us from a distance and then abruptly turn and hop away the moment we get too close. They're wary, shy creatures. But today, they seem reluctant to leave their juicy pastures. Or maybe they're just growing more accustomed to our visits (see Figure 2.1).

When two of the girls notice a rather large joey still in its mother's pouch, they edge farther down into the gully to get a better look. They've never managed to get so close before. An enormous buck is watching the girls intently, but he doesn't flinch. Following his lead, the mob stays put. The girls eventually stop as well. For a suspended moment, girls and kangaroos face each other off at close quarters.

In *Wild Dog Dreaming: Love and Extinction*, Deborah Bird Rose (2011) questions Levinas's narrow framing of ethics as an exclusively human ability to recognise our mutual vulnerabilities in the face-to-face encounter with another. She interprets Levinas's insistence that ethics precedes, rather than resides in the self, to infer that it is ethics 'which calls us into relationship' (Rose 2011, p. 29). At a time in which it is no longer possible to deny that our tenuous survival is inextricably linked with the world's other lifeforms, she questions why we would exclude

Figure 2.1 Child-kangaroo encounters.
Author's (Affrica Taylor's) photograph

other animals from a relational ethics. Paying homage to environmental feminist Val Plumwood, she and her colleagues urge us to 'resituate the human within the environment, and resituate nonhumans within cultural and ethical domains' (Rose *et al.* 2012, p. 3, see also Rose 2015).

When deliberating on the ethical possibilities that open up in the moment of kangaroo-child face-to-face encounters such as this, we have lots of unanswerable questions. What stirs kangaroo and child curiosities about each other? What compels them to observe each other so closely? What do they recognise – or not – through exchanging gazes? Do they apprehend, in some way, their mutual vulnerabilities? We wonder if these fleeting encounters contain an element of mutual calling. And, if so, could they be the start of an ethical relationship?

For the first time, it's the girls, not the kangaroos, who are the ones to turn and run away – when the intensity of the moment becomes too much, when the zones of proximity are breached. The imposing buck is steadfast, resolutely guarding his mob. He doesn't take his eyes or his ears off the girls. Ears forward, in sync, all the adult kangaroos study their retreat. The girls laugh with nervous excitement as they run back. They look pleased with themselves. Stimulated by the up-close physicality of kangaroo bodies? Perhaps imagining a chase?

In relation to her own entanglement in multispecies lifeworlds that we can never fully understand or control, Lesley Instone (2015) takes up the challenge of risking attachments in the Anthropocene. She argues that the process of risking attachments holds more possibilities than dangers:

> Such a stance means different ways of thinking and doing that connect us as one among the many actors and places that enact the world' she says, while encouraging us to see that 'embracing our attachments and embeddedness in complex networks offers hope rather than menace.
>
> (Instone 2015, p. 31)

Is it risk or being on the edge of risking attachment that moves the children as they laugh and run away? And what of the kangaroos? Are they, too, sensing risk or risking attachment as they stand their ground – longer than usual this time, before taking flight? In taking the risk of being closer to each other this time, do children and kangaroos also sense the inevitability of their entanglements with each other in some way? Is this risk they take in this extended moment a small example of the modest kind of 'multispecies achievement' that van Dooren and Rose (2012, p. 3) speak of as the necessary precursor to an ethics of multispecies conviviality?

We make our way back up the hill towards the campus buildings and onto the concrete path that leads to the early childhood centre. The girls hop ahead. Later, one of them draws a picture of two small girls standing next to an enormous kangaroo and holding hands with the joey in the kangaroo's pouch. She explains to me that she and her friend and the joey are holding hands 'because we are all close'. They do seem to have a familial connection.

There is always some kind of contagion at work in Haraway's (2008) account-ings of what happens 'when species meet'. Again, it has something to do with the play of bodies, where relating precedes identity. Where the flesh-and-blood meeting infects a lateral (rather than a reproductive lineage and descent) kind of kinship (Haraway 2015). The children seem so drawn to the joeys. Could it be that this up-close encounter with an actual joey might have infected them with a sense of kinship that exceeds both the human-centric and heteronormative notion of biological kinship, as well as the culturally sanctioned attraction to the 'cute' young of other species so heavily promoted in children's popular culture? Might this kinship gesture between two young girls and a young kangaroo foreshadow the emergence of a multispecies ethics of belonging?

We've been watching the film version of Dot and the Kangaroo *lately. It shows how a kind kangaroo ensconces a lost human child in her pouch and hops off to deliver her home. As well as rescuing Dot, Kangaroo secures the child's sympa-thies for the bush animals that are hunted by the 'cruel white people', and Dot promises to become an 'improved human'. Today, one of the girls wore her kan-garoo hoody jacket (complete with ears) especially for the bush walk. Perhaps the Dot story has stirred her imagination about the possibilities of a new kind of hybrid girl-kangaroo identity. As we approach the mob, her best friend encour-ages her to go ahead to mediate the encounter. She seems willing and brave.*

Do the children really think that the kangaroos would mis-recognise a hop-ping girl in a kangaroo suit as one of them? Does her furry jacket make her feel kangaroo-like as she hops confidently forward? Does she see the kangaroos differently in her suit? Does she see them as kin? And what do the kangaroos make of her?

Acknowledging the influence of Barbara Noske's work on her thinking, in particular Noske's insistence that we must recognise 'the "otherworldly" subject status' of other animals, Haraway (2004, p. 143) asks, 'What is intersubjectivity between radically different kinds of subjects?' In her own work, she has consistently called for 'otherwordly conversations' with animals that refuse 'autonomisation of the self', as well as 'objectification of the other' (Haraway 2004, p. 144). Haraway takes on the daunting task of trying to articulate the ways in which we are constituted through our relations with other species. She tries to do this in ways that resist reducing these others into known objects of interest to us (as all-knowing human subjects). Could there be a similar realignment of subject-object multispecies relations at play in this particular child-kangaroo encounter? As the kangaroo girl breaks with the human group and hops towards the mob, as she bodily enacts becoming kangaroo, is she called into a trans-species mode of intersubjectivity? Is she entering into a kind of bodily enacted 'otherworldly conversation'? Are these the kinds of conversations we all need to be open to if we are to recognise other species as agentic subjects whose life stories are co-shaping ours – particularly when our worlds collide?

Children and deer: entangled inheritances

As regular grazers of suburban parks, verges, and front yard gardens, deer co-shape North American urban lives. They are amongst the most ubiquitous North American animals and, like kangaroos in Australia, are a Canadian emblem, strongly defining settler Canadian identity. For instance, in the Canadian West, a wapiti (highly evolved old deer), alongside a bighorn sheep, supports the shield of British Columbia's coat of arms. In the Prairies, a royal lion and a white-tailed deer are featured in Saskatchewan's coat of arm. In central Canada, a moose, a deer, and a black bear are highlighted in Ontario's coat of arms. In Eastern Canada, two white-tailed deer and an Atlantic salmon play a significant part in supporting the shield in New Brunswick's coat of arms (Government of Canada 2015). Granted by the British Empire in the 19th century, these *white* Canadian symbols have absorbed the deer (and other featured native animals) into colonialist versions of the land that presume unity of settlers and indigenous peoples under a colonial set of rules, ideas, and illusions.

Today's complex inheritances of children and deer are unavoidably entangled with these settler colonial relations. Like the kangaroo in Australia, the deer is not only a Canadian emblem but also a figure that secures settler children's connection to the 'new' land by erasing indigenous peoples' relations with forest creatures. Settler narratives of belonging and identity depicted in many North American children's stories and popular culture are one way in which this connection is fortified.

Deer characters in children's popular culture

Although there are many North American children's deer stories, the iconic Walt Disney film *Bambi* (1942), based on Salten's (1998 [1928]) novel *Bambi: A Life in the Woods*, is perhaps one with which most anglophone children are familiar. Since the mid-20th century, this film has encouraged generations of children to affectionately identify with deer. *Bambi* tells the life story of a deer growing up in an idyllic North American forest. As Taylor (2013, pp. 19–21) details in her analysis of the Disney film, it works to associate an explicitly white-settler romantic imaginary of pristine, unpeopled wilderness with natural innocence and beauty and, at the same time, to erase any sense of indigenous peoples from the land.

As a typical Disney film, the story begins with a doe giving birth to a fawn named Bambi, whose destiny is to protect the innocence and beauty of the forest and its animals. Bambi grows up in an uncannily bucolic forest that he shares with a rabbit named Thumper, a young skunk named Flower, and a fawn named Faline. Like his father, Bambi eventually becomes 'Great Prince of the Forest'. The first part of the film features Bambi's mother teaching and cautioning Bambi about the dangers of life in the forest – namely the ever-lurking yet never-seen evil human hunters. The second part of the film portrays Bambi's life with Faline, including the birth of their twin fawns. There are key moments in the plot that make Bambi a 'storyline of conservation at its very heart' (Whitley 2008, p. 64). The first is when Bambi's mother is shot and killed by a hunter when she is teaching Bambi how to find food. A second pivotal moment is when Bambi is awakened by the smell of smoke, and his father warns him of a wildfire, lit by hunters. A third one is when Bambi finds Faline cornered by vicious hunting dogs, belonging to the hunters, which he manages to ward off.

Although a conservation story, which condemns the killing of innocent wildlife and the destruction of pristine wilderness, *Bambi* is nevertheless narrated through a resolutely settler colonial gaze that renders invisible indigenous human presences and relationships with the land and animals. Thus, *Bambi* produces 'a very selective vision of human interaction with the landscape, eliminating any sense of an evolving interdependency between landscape and human activity' (Whitley 2008, p. 67). Moreover, the two human practices that are portrayed as 'evil' in the name of conservation – burning the land and hunting – are both traditional indigenous practices that were a central part of the life and death co-existence and mutual flourishing of the multispecies forest ecosystem.

Fraught settler – deer relations

The residual colonial gaze of *Bambi*, as a mid-20th-century, white-settler conservation fantasy, must be framed within the broader suite of settler struggles to establish their dominance within what they called this 'new' land and, more specifically, within the fraught politics of settler-deer relations. White-tailed and mule deer have been evolving in North America for about four million years (Nelson 1998), and their meat has been a staple of the First Nations' diet for thousands of years. Because the first Europeans to come to North America also depended on deer meat to survive, deer quickly became an important link between First Nations and the white settlers. It was not long, however, before the European relationship to deer shifted from one of sustenance and survival to one of capital accumulation and empire building. European settlers hunted them to provision trading posts and export their hides, and by the beginning of the 20th century, deer, along with many other species, were at risk of being wiped out across North America (Cambronne 2013).

By the beginning of the 21st century, tables had turned, and the future for deer populations in North America had been secured. This was due to the restocking of deer species, prompted by conservationists as well as market pressures, and sustained by warmer temperatures, a decline in predators, and their own remarkable ability to adapt to living in a diverse range of settler-altered environments. Today, deer are plentiful and can be seen grazing across a wide belt of North American forests, agricultural lands, and urban green spaces. The economic incentives for preserving and growing a profitable deer 'industry' cannot be underestimated. Since World War II, recreational hunting and tourism have boosted the value of deer in the North American economy to millions of dollars (Cambronne 2013).

All of these historical and cultural legacies produce a strange mix of ambivalent settler-deer relations. The sentimental and affectionate attachment to deer that *Bambi* has instilled in North American audiences, young and old, still remains. Deer are also still important to North American economies. However, the increasing trend for deer to flourish in urban environments, to the point of overabundance, is now perceived as an inconvenience, and even as a threat by many city dwellers.

Deer in Victoria

The city of Victoria, on Vancouver Island, is a case in point. Here, deer thrive and abound in the lush urban forests, parklands, and gardens, so much so that a virtual war has been declared on deer. Media headlines reflect, if not promote, such local sentiments: 'No Clear Solution to Victoria Region's Deer Problem' (Palmer 2012); 'Victoria Overrun by Deer as Some Locals Propose Cull' (Clarke 2011); 'Haro Woods Deer a Hazard for Drivers' (Paterson 2014). Plentiful food, relatively mild winter conditions, a lack of predators such as cougars and wolves, and increasing human encroachment into deer habitat have resulted in high numbers of deer living within and around Victoria's urban precincts. Biologists estimate that Vancouver Island is home to 75 percent of British Columbia's mule deer, or approximately 86,000 deer (Clarke 2011), and the provincial and regional governments have declared that the total number of deer in Victoria is unsustainable. This is expressed in highly human-centric terms. For instance, official rhetoric describes deer as having exceeded the 'cultural carrying capacity', or the 'maximum number of ungulates that can coexist compatibly with local human populations', which is determined by 'the sensitivity of the local human population to the presence of animals' (Hesse 2010, p. 7).

Some environmentalists also argue that the overabundance of deer poses a threat to the deer population itself, which has to compete for food, as well as to the forests' vegetation and ecosystems (Côté *et al.* 2004). The deer-human controversy that drives the debate in the city, however, is framed primarily around the inconveniences that deer present to humans, including as carriers of potential diseases, destroyers of gardens, attackers of pets, and aggressors during rutting season, not to mention the inconvenience and costs of deer-proof fencing, destruction of agricultural land, and car-deer collisions. The financial burden of urban deer 'damage' drive the majority of these concerns, as evidenced by cost estimates in the city budget (Denis 2013). As in Canberra, ecological rationales, underpinned by economic motives, have led to highly controversial culls in some of Victoria's neighbourhoods.

With many parallels to the fraught settler-kangaroo relations in Australia and the dilemmas of human-kangaroo cohabitation in Canberra, these deer-human stories also make us wonder about the ethics of multispecies cohabitation in contemporary urban settings. Is it possible to coexist with deer in the city in respectful and mutually sustainable ways? How might we help children to negotiate the ambivalent legacies of settler-deer relations that they have inherited? In the following section, we narrate the real-life encounters of children and deer in an early childhood education centre in Victoria. Educators and children in this centre are grappling with the question of 'how to live well with local deer, mindful of the many interconnected stories and experiences that humans and deer bring into this shared space and which continue to co-shape each other's lives and futures'. They take daily walks through the forest that is adjacent to the centre and home to a large herd of mule deer.

Child-deer encounters

The two girls walking at the front of the line come to an abrupt halt on the forest path. One of them turns around and whispers an urgent command to others behind her: 'The deer are watching us, stop!' There is a palpable sense that we are visitors in someone else's place. As always, the deer are alerted to our presence in the forest well before we see them. Our human scent has given us away. Deer have an amazing sense of smell.

To the children's delight, the deer also visit the centre regularly, coming right up to the chain-link fence that delineates their 'natural' forest territory from the human 'cultural' precinct. Through the wire, the children can get a really good view of the deer. Much to everyone's surprise, two deer have started to bed down in the grass beside the fence, and they are still there in the early mornings. We all wonder why these deer haven't chosen to hunker down in the deep forest where they aren't so visible or exposed to predators. They must feel safe in this liminal natureculture zone.

As the deer and the children encounter each other, new multispecies stories emerge, new response-abilities are enacted, and new possibilities are generated for learning how to shape and reshape worlds together. Might it be, as Rose (2014, para. 3) suggests in regard to other multispecies relations, that the children and the deer in this forest are creating a convivial bond of cohabitation 'in a time when climate and other pressures are putting stresses on everyone's capacity to survive'?

We are particularly interested in the complex ways in which the chain fence mediates the relationship between the deer and the children. As Lesley Instone (2010, p. 93) reminds us, 'the fence is an arresting delineation of native/non-native, introduced/indigenous, colonial/postcolonial'; it reflects how material and discursive practices of 'neat/messy, familiar/unfamiliar, accessible/inaccessible' do the work of boundary maintenance. Certainly, this chain fence can be seen to be doing boundary maintenance work by materially and semiotically demarcating (and protecting) the children's 'civilised' manicured playground territory from the deer's 'wild' territories of tall native grasses and overgrown pines. On the other hand, Instone invites us to think with the fence in more complex ways. As she puts it, 'The fence is a dynamic space of contestation and interaction that activates all manner of work', creating 'a line of communication, not just a division' (Instone 2010, p. 97). The fence that physically prevents the deer from entering the children's territory also allows for a line of communication between them. It allows deer to settle in closely without being disturbed. It allows both children and deer the opportunity for physical proximity and habituation to each other's differences. This fence has become a space of conjunction, of possibility, and of connectivity. In Instone's (2010, p. 111) terms, it has facilitated a new kind of attunement, risk taking, and attachment between children and deer that might 'nudge' them 'towards an ethics of co-transformation'.

Back in the forest, the doe and two fawns that arrested the children on the forest path slowly walk away. The two lead girls want to find out where they are going. They ask me (Veronica) to come with them. Cautiously avoiding approaching too quickly, and in great anticipation, we walk after the doe and fawns. In a gap between sapling pines and English ivy, the deer stop and turn around, their ears, tipping back and forth to better hear us. We freeze. The deer and the girls stare at each other. The deer continue walking slowly for another metre or so. The girls follow. Once again, the deer stop and look around at us for another few seconds. Again, we freeze in total silence. Without seeming too concerned, the deer resume walking. A kind of playful rhythm has been established. It's starting to feel like a game of statues. There are several more advances, freezes, and exchanges of gazes before the deer reach a thick part of the forest that precludes us. But through the thickets, we can just make out that the deer are still turning around to see if we are following.

In *When Species Meet*, Haraway (2008, p. 232) writes about interspecies play and notes that such 'play builds powerful affectionate and cognitive bonds between partners'. Through engaging in it, very different partners get to know and interpret each other's modes of movement and communication, tentatively learning 'the sharable meanings and practices of social bonding and territory identification' across categorical boundaries (Haraway 2008, p. 233). Play in the interspecies contact zone, Haraway continues, 'brings us into the open where purposes and functions are given a rest' (p. 237), and in this unknowable opening between species, partners must allow themselves up 'to inhabit a muddle' (Haraway 2008, p. 238). Perhaps it is not too farfetched to view this exchange between the deer and the girls as a 'noninnocent playful investigation', a slightly halting, uncoordinated, and muddled kind of investigative play, through which both partners were nevertheless participating in 'the reinvention of the world' (Haraway 2008, p. 237).

A couple of weeks after the exchange, the children read Kate and Pippin: An Unlikely Love Story *(Springett 2011). This is a photographic essay that documents a real-life interspecies relationship that developed between a Great Dane (Kate) and an orphaned fawn (Pippin) on Vancouver Island. From their first meeting, when the abandoned fawn has just been 'rescued' from the forest into the Great Dane's family home to be fed and cared for, Kate showed a great maternal interest in her. She nuzzled and licked the fawn in the same way that her deer mother might have done. This marked the start of an unlikely but enduring interspecies relationship, which the author notes changed both dog and deer forever. Even though Pippin was eventually returned to the forest to live an independent life, she has retained her bond with Kate, the Great Dane, whom she visits regularly to this day. The deer and the dog have become life companions, forging a new kind of cross-species kinship.*

Upon reading the book, the girls recall their own encounter with the deer in the forest, comparing how the deer made space for them to follow and how the Great Dane made space for the fawn. Excitedly, they plan what they might do the next time they face the deer in the forest. They are thoughtfully speculative and very open to the question 'what if?'

Although this book received a lot of critical media attention for its apparent sentimentality, anthropomorphism, and for sanctioning the domestication of wild animals, the children found it quite compelling. It seemed very easy for them to relate this story of cross-species love and care to their own desires to form close relationships with deer. We wonder how much this is simply a reflection of their cultural conditioning to identify with highly romanticised children's stories about anthropomorphised animals – as exemplified in texts like *Bambi* – or how much it is about something else entirely. Discussing ways of finding hope for multispecies cohabitation in this time of anthropogenic extinctions, Rose (2016) considers hope as being 'life's desire for more life' and a sense of being interwoven within the fabric of mutually dependent but very different lives. Might stories such as this one, that document how different species seek attachment rather than walk away, convey a hopeful sense of 'life's desire for more life' to the children? Might they give the children a sense that multispecies relations are central to the interwoven fabric of life?

A few years ago, children at this early childhood centre began a project of crafting deer that has become an ongoing work-in-progress. They began by constructing a twig deer, and later on they resurrected two wire deer found in the trash after neighbours cleaned up their Christmas decorations from their front yard.

The wire deer were carefully wrapped with fabric that the children dyed with forest plants and other natural dyes. A year later, a different group of children moved the fabric-maché deer into their studio, where they were experimenting with paint. The next year's cohort of children collaborated to feel the deer. For the last three years, subsequent groups of children have crafted and recrafted this wire-trashed-fabric-machéd-paint-decorated-felted deer – and in the process, they have created a space in which to think with and plan a close encounter with the forest deer.

Today, the children have taken the crafted deer with them on their walk (see Figure 2.2), in the hope that it will attract the live deer in the forest to come by to say hello . . . They strategically position it in different locations and wait, but the forest deer do not show up. Excitement rises when they finally spot a doe and a buck off in the distance. Although the forest deer are far away, the children convince themselves and each other that the deer have made themselves visible because they're curious about the crafted deer. They wait patiently, holding on to hope that their lure will draw the deer closer.

These ongoing deer-child encounters are characterised by mutual curiosity, attentiveness, uncertainty, inventiveness, compromise, risk, generosity, and connection. From the children's part, they are driven by a strong desire to become closer and closer to the deer and to include them within their ambit of common world belongings. Might these small acts of inclusion on the part of the children, however faltering and imperfect, be viewed as multispecies achievements, as opportunities for a future of more congenial multispecies urban cohabitations?

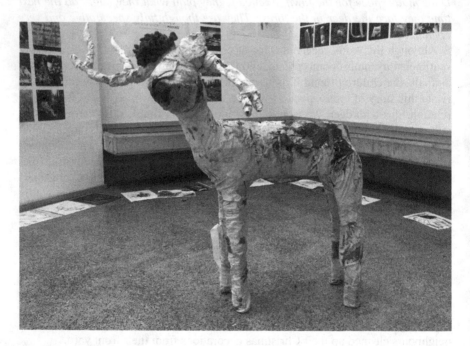

Figure 2.2 The crafted deer.

Author's (Veronica Pacini-Ketchabaw's) photograph

More specifically, do the children's thoughtful and deliberate attempts to get close to the deer provide other pathways for human-deer relations in the city, beyond municipal discussions about inconvenience and financial costs?

A multispecies ethics of conviviality

The child-kangaroo and child-deer stories that we have recounted in this chapter are geo-historically specific, thus deliberately idiosyncratic and parochial. We have not offered them as a way of representing the normative relationships of all children and wildlife everywhere, but to detail the very distinctive ways in which particular children's lives are already entangled with those of particular wild animals, even before they meet. The parochial nature of these stories illustrates the ways in which the macro-political context of anthropogenic environmental damage – closely associated with settler colonialism in these two cases – materialises in quite specific ways on the micro-political grounds of local urban neighbourhoods – where actual children and actual wild animals encounter each other in the course of their everyday lives.

It is these lively, emplaced, and embodied encounters that we are particularly interested in. They testify to van Dooren and Rose's (2012, p. 16) insistence that cities are never exclusively human spaces, but always co-constituted through the generative life stories and interactions of many different species. But perhaps more importantly, they throw light on their driving ethical concern, which is to find new and convivial ways of sharing urban spaces with wildlife in this time of compelling environmental challenges (van Dooren and Rose 2012, p. 2).

For van Dooren and Rose (2012, p. 17), an ethics of conviviality means *really* sharing urban spaces and creating a 'genuinely inclusive multispecies city' that 'provides a space for the flourishing of as many different forms of life as possible'. Extending Fincher and Iveson's descriptions of a convivial, culturally diverse city, they explain that multispecies urban conviviality is not about presuming to share an identity with other species or to reduce very different experiences of living in the same place to one thing. However, it does require us to at least temporarily identify with and/or make some kind of connection with other species by being attentive their presence and to their distinctive ways of being in this shared place (van Dooren and Rose 2012, p. 17). In our multispecies research, we were particularly struck by the children's drive, desire, and capacity to do precisely this – to share a mutual hyperawareness of each other's presences with the kangaroos and deer, to identify with them (particularly the joeys and fawns), to try to connect with them, and to pay close attention to kangaroo and deer embodied modes of being.

We believe there is much to learn from attending to the details of child-wildlife encounters such as these. Although quite fleeting and seemingly minor events, they are examples of the small multispecies achievements that occur on a daily basis but are rarely noted because they are beyond the human-centric, and adult-centric, framings of significant relations. As restorying strategies (Nxumalo 2015), these vignettes of fleeting child-kangaroo and child-deer encounters open up a space for reimaging an inclusive multispecies city. They show us that a curious fascination with the incommensurable differences of other lifeforms can drive an ethics of conviviality. It is really quite simple. If, like these children, we are genuinely interested in the lives of others with whom we share a 'common world', we are well on the way to enacting a recuperative ethics of multispecies conviviality and belonging in the city.

Note

1 For a detailed analysis of Pedley's *Dot and the Kangaroo* and discussion about the figure of the kangaroo in the cultural politics of Australian nationalism, see Taylor (2014).

References

Australian Capital Territory, Territory and Municipal Services, 2010. *ACT kangaroo management plan*. Available from: www.accesscanberra.act.gov.au/app/answers/detail/a_id/736/~/kangaroo-management [Accessed 9 February 2018].

Australian Government Department of Foreign Affairs and Trade, 2013. *About Australia: Coat of arms*. Available from: www.dfat.gov/facts/coat_of_arms.html [Accessed 15 October 2013].

Bambi, 1942. Animated film. Produced by W. Disney Productions, directed by D. Hand. USA: RKO Radio Pictures.

Cambronne, A., 2013. *Deerland: America's hunt for ecological balance and the essence of wildness*. Guilford: Lyons Press.

Clarke, B., 2011. Victoria overrun by deer as some locals propose cull. *The Globe and Mail*, 1 November 2011. Available from: www.theglobeandmail.com/news/british-columbia/victoria-overrun-by-deer-as-some-locals-propose-cull/article4199893/ [Accessed 9 February 2018].

Côté, S., Rooney, T.P., Tremblay, J.-P., Dussault, C. and Waller, D.M., 2004. Ecological impacts of deer overabundance. *Annual Review of Ecology, Evolution, and Systematics*, 35, 113–147. Available from: https://pdfs.semanticscholar.org/3f07/6d402f0a44cbd373 32935ec912b9574e3375.pdf [Accessed 17 February 2018].

Denis, A., 2013. *City of Victoria deer management plan, cost estimates*. Available from: www.victoria.ca/EN/main/city/city-budget/2013-budget-information.html [Archived].

Flannery, T., 2008. *Chasing kangaroos: A continent, a scientist and the world's most extraordinary creature*. 2nd ed. New York: Grove Press.

Franklin, A., 2006. *Animal nation: The true story of animals and Australia*. Sydney: UNSW Press.

Gibson, K., Rose, D.B. and Fincher, R., 2015. *Manifesto for living in the Anthropocene*. Brooklyn: Punctum.

Government of Canada, 2015. *New Brunswick's provincial symbols: Coat of arms*. Available from: http://canada.pch.gc.ca/eng/1444070816858#a3 [Accessed 9 February 2018].

Haraway, D.J., 2004. Otherworldly conversations; terran topics; local terms. *In*: *The Haraway reader*. New York: Routledge, 125–150.

———., 2008. *When species meet*. Minneapolis: University of Minnesota Press.

———., 2015. Anthropocene, capitalocene, plantationocene, chthulucene: Making kin. *Environmental Humanities*, 6, 159–165. Available from: http://environmentalhumanities.org/arch/vol6/6.7.pdf [Accessed 9 February 2018].

Hesse, G., 2010. *British Columbia urban ungulate conflict analysis summary report for municipalities.* British Columbia Ministry of the Environment. Available from: www2. gov.bc.ca/assets/gov/environment/plants-animals-and-ecosystems/wildlife-wildlife-habitat/staying-safe-around-wildlife/urbanungulatesconflictanalysisfinaljuly5–2010. pdf [Accessed 9 February 2018].

Instone, L., 2010. Encountering native grasslands: Matters of concern in an urban park. *Australian Humanities Review*, 49, 91–117.

———., 2015. Risking attachments in the Anthropocene. *In*: K. Gibson, D.B. Rose and R. Fincher, eds. *Ethics for the Anthropocene.* New York: Punctum Books, 29–36.

Lorimer, J., 2015. *Wildlife in the Anthropocene: Conservation after nature.* Minneapolis: University of Minnesota Press.

Nelson, M.E., 1998. Development of migratory behaviour in northern white-tailed deer. *Canadian Journal of Zoology*, 76, 426–432.

Nxumalo, F., 2015. Forest stories: Re-storying children's encounters with 'natural' places in early childhood education. *In*: V. Pacini-Ketchabaw and A. Taylor, eds. *Unsettling the colonialist places and spaces of early childhood education.* New York: Routledge, 32–63.

Palmer, D., 2012. No clear solution to Victoria region's deer problem. *Victoria News*, 4 September 2012. Available from: www.vicnews.com/news/168538686.html [Accessed 9 February 2018].

Paterson, T., 2014. Haro Woods deer a hazard for drivers. *Saanich News*, 26 November 2014. Available from: https://issuu.com/blackpress/docs/i20141126051213541/4 [Accessed 9 February 2018].

Pedley, E.C., 1997. *Dot and the kangaroo.* Sydney: Angus and Robertson. Reproduction of original Angus and Robertson 1906 print edition prepared by Sydney University Library, 1997. Available from: http://adc.library.usyd.edu.au/index.jsp?page=home& database=ozlit [Accessed 9 February 2018].

Rose, D.B., 2011. *Wild dog dreaming: Love and extinction.* Charlottesville: University of Virginia Press.

———., 2014. *Apologising to dingoes* [blog post]. Available from: http://deborahbirdrose. com/tag/barry-lopez/ [Accessed 9 February 2018].

———., 2015. The ecological humanities. *In*: K. Gibson, D.B. Rose and R. Fincher, eds. *Manifesto for living in the Anthropocene.* New York: Punctum Books, 1–6.

———., 2016. *Hope is the way of the world.* Available from: http://deborahbirdrose. com/2016/05/31/hope-is-the-way-of-the-world/ [Accessed 9 February 2018].

Rose, D.B., van Dooren, T., Chrulew, M., Cooke, S., Kearnes, M. and O'Gorman, E., 2012. Thinking through the environment, unsettling the humanities. *Environmental Humanities*, 1, 1–5. Available from: https://environmentalhumanities.org/arch/vol1/EH1.1.pdf [Accessed 17 February 2018].

Salten, F., 1998. *Bambi: A life in the woods.* New York: Simon and Schuster (Original work published 1928).

SEARCH, 2013. Unearthing Australia's climate history. *South Eastern Australia Recent Climate History*. Available from: http://climatehistory.com.au/ [Accessed 9 February 2018].

Simons, J., 2013. *Kangaroo.* London: Reaktion Books.

Springett, M., 2011. *Kate & Pippin: An unlikely love story.* Toronto: Penguin.

Taylor, A., 2013. *Reconfiguring the natures of childhood.* London: Routledge.

———., 2014. Settler children, kangaroos, and the cultural politics of Australian national belonging. *Global Studies of Childhood*, 4(3), 169–182.

van Dooren, T., 2014. *Flight ways: Life and loss at the edge of extinction*. New York: Columbia University Press.
van Dooren, T. and Rose, D.B., 2012. Storied places in a multispecies city. *Humanimalia: A Journal of Human/Animal Interface Studies*, 3(2), 1–27.
Whitley, D., 2008. *The idea of nature in Disney animation*. Farnham: Ashgate.

3 Children, ants, and worms
An environmental ethics of mutual vulnerability

An environmental ethics of vulnerability in which humans are vulnerable to living and nonliving earth processes [is] concerned not only with past and present damage but also with future unknowable vulnerability to harm.

(Hird 2013b, p. 107)

Interspecies vulnerabilities

The thing that worries us most about an uncritical engagement with the Anthropocene is that it can easily be taken as a reconfirmation of human exceptionalism and a concomitant excuse to redouble efforts to further/better 'engineer' the environment to serve human interests. The dissociative sense of omnipotence and impunity that accompanies responses that assume that humans can improve on and/or rescue nature seems to be part and parcel of the original problem, not the solution to it. Along with a number of other feminist scholars who are also wary of the potential triumphalism of the Anthropocene (Haraway *et al.* 2015, Stengers 2016), we respond by trying to better understand the ways in which human pasts, presents, and futures are inextricably bound up with those of other earthly beings and forces. Moreover, as the opening quote from Hird (2013b) implies, rather than assuming that we somehow transcend the fates and futures of other lifeforms and environmental processes damaged by our actions, we prefer to look at how our own lives, fates, and futures, as well as the lives, fates, and futures of other species, are mutually implicated and thus mutually vulnerable.

In this chapter, we think through the ethical possibilities of interspecies vulnerability by focusing upon a series of seemingly insignificant everyday encounters between seemingly minor actors – between young children and worms in a wet Canadian forest and between young children and ants in a dry Australian bushland. Taken from our multispecies ethnographic research, these encounters illuminate some of the ways in which we are vulnerable to other species and they are vulnerable to us. They make us aware of the vulnerability of ants and worms to children's poking fingers and wayward feet, but also prompt us to reconsider how we are also vulnerable to these small but indispensable subterranean creatures. This focus upon interspecies vulnerability helps us to rethink our place in the world as just one of many earth shapers and makers. It tempers the exceptionalist imaginary that the world is all about us and reminds us that other lifeforms and earth processes create the conditions of possibility for our human lives. Moreover, they do so in ways that we do not always fully acknowledge or comprehend.

We are particularly interested in thinking through the mutual vulnerabilities of these ant-worm-child encounters in the context of our ecologically precarious futures. Our awareness of this precarity has been sharpened by the Anthropocene debates and has spurred our interest in exploring ethical responses that do not rehearse the conceits of assumed human exceptionalism and exclusive agency.

Even though these child-worm-ant interspecies encounters are seemingly insignificant, small, and ordinary events, we situate the relational possibilities that flow from them within the huge ethical questions posed by the Anthropocene. We also borrow heavily from Myra Hird's (2010, 2012, 2013a, 2013b) insightful microbial sociologies to help us think these connections through. Based on her tutelage in Lyn Margolis's Massachusetts microbiology laboratory and her own passionate engagement with bacteria in Canada's waste landfill sites (2013a), Hird's (2010) microontologies have given us new ways to consider the potent agency and worldly significance of the subterranean worlds of worms and ants.

Hird (2013a) notes that while humanism might predispose us to care selectively about some other animals that are 'big like us' and acknowledge that they have an auxiliary role to play, along with us protagonists, in 'the greatest show on earth', it usually fails to incite our interest in smaller lifeforms. She points out that when engaging with and thinking through bacteria, the less glamorous but nevertheless essential and real action takes place backstage. Human lives and actions might be big and spectacular, but it is the backstage metabolising work of microbes and bacteria that creates the conditions of possibility for these front-of-stage lives and actions (Hird 2013a). We could similarly point to the essential life-supporting work that is going on underground, where quadrillions of worms and ants are engineering the show that we, as humans, are simultaneously putting on and witnessing on the surface of the earth.

Thinking through the microbial world, Hird (2010, p. 36) suggests that we need to disengage ourselves from humanism's 'profoundly myopic' ontologies and ethics. She points out that even when we seek to include the nonhuman animal in the ethical realm from a humanist position, we still 'pivot on a comparison between the humans and [the] animal' (Hird 2010, p. 36). As she explains, this is because humanist ontologies can only ever produce an 'Other-ethics based on face-to-face interaction', an ethics that constantly defaults to the human-as-type and relies on some kind of anthropomorphic recognition of ourselves in the other animal (Hird 2012, p. 262). In other words, while it is possible to extend a sense of ethical responsibility to the animal with appealing big eyes (like ours) and a cute face (like ours) within a fundamentally humanist ethical schema, it is not so easy to include those lifeforms that have no face at all – like bacteria and earthworms. Hird (2012, p. 262) proposes a radically different kind of ethics, one that does not rely on the human as the default extender of ethical care, and which she calls an ethics of vulnerability. Within this 'inhuman' form of ethics, we are called into account for our vulnerability alone, not for our largesse and capacity to care for others. The agency is completely reversed when we become beholden to the myriad of micro lifeforms we rarely see, let alone acknowledge, and yet which sustain the lives of all large animal species, including our own.

In an extension of these ideas and in stronger recognition of the Anthropocene, Hird (2013b, p. 105) writes about an 'environmental ethics of vulnerability' that is 'sensitive to human and nonhuman asymmetrical vulnerability to an unknowable future'. In other words, this environmental ethics of vulnerability is one that recognises that although all species have inherited and are co-implicated in the environmental uncertainty and mess we now find ourselves in, it might well be that we humans are amongst the most vulnerable species of all.

Our efforts to develop a multispecies ethics of environmental vulnerability also draw directly from our ethnographic observations of the relationalities, interdependencies, and encounters between children and animals in their local common world environments. These observations are longitudinal, situated, immersed, relationship based, and affect-attuned (Kirksey and Helmreich 2010; Ogden *et al.* 2013). They recognise that mutual vulnerabilities emerge, not only in singular moments of multispecies encounter, for instance, when children, ants, and worms meet, but that our multispecies lives and futures are already co-implicated simply because they are interdependent and entangled in mutually constituted and now fundamentally damaged worlds.

In order to highlight these interdependencies and entanglements, we attend to what Anna Tsing (2013) refers to as 'more than human sociality', in which all of the actors learn about each other *in action* and researchers become companion participants in sticky webs of connection engaging in experimental and inventive practices. Using inventive and experimental co-implicated research practices, this chapter contemplates new kinds of observational skills that are akin to what Latour (2004, p. 206) calls 'learning to be affected'. Learning to be affected requires us to develop more-than-cognitive modes of attention – to become attuned to the multifarious ways that human and nonhuman bodies are moved, disconcerted, and enlivened through their common world encounters. In our multispecies research, this not only means paying attention to what the children are saying and doing and how their bodies are being moved, affected, and enlivened by the small subterranean animals they encounter but also to the movements and actions of the worms, ants, water, rain boots, fingers, sticks, rocks, mud, pebbles, and dust – to the whole assemblage of the child-worm-ant encounters. We push ourselves to learn to be affected by and think with all of the actors – in particular by and with the children's, the ants', and the worms' bodies, movements, disconcertments, and preferences – even though, as Tsing (2013) points out, and as we acknowledge, this is extremely hard work.

The hard work is not just about noticing the multitude of things that are going on for human and nonhuman others. It is also about noticing what human and nonhuman others notice (Tsing 2013). Of course, this is never fully achievable. The limits to human intentionalities, observations, and knowings are premised in multispecies ethnographies, precisely because they acknowledge that humans are not the only ones exercising agency and not the only animals noticing, observing, acting, knowing, affecting, and being affected. We recognise that our efforts to notice by being there and being interested (Hird 2013a), by staying attuned and learning to be affected (Latour 2004), by asking questions and paying attention to the temporal and the transitive (Lorimer 2010), by using images as processes of 'collaborating and moving *with* the world' (Kind 2013, p. 429), and by pushing ourselves to think through and with a multiplicity of more than human actors (Whatmore 2013) are always delimited by the partiality of our human apprehensions and ambivalences.

The following sections are structured around a range of fragmented elements, including vignettes of child-worm and child-ant encounters, reflections on ant and worm sciences, and ethical ponderings prompted by the work of feminist scholars working in the environmental humanities. Challenging the humanist research conventions of the all-knowing-human-actor-as-researcher, we resist responding to the things we do notice during these multispecies encounters as objectified findings to be analysed. Instead, we deliberately ask questions, open up possibilities, and remain curious, but also admit that there is always more going on than we can ever know. Our intention is to evoke and provoke, rather than represent or explain, what might be going on in those encounters (Lorimer 2010).

Thinking with children and worms

When we think about the history of earthworms, we are reminded of our vulnerabilities, particularly when we recall that they predate us by 600 million years, and even though we now live with them, they will likely outlive us (Schwartz 2012). These invertebrate animals, once considered lower organisms, are deeply entangled with other species' lives, and, more poignantly, are active members of the history of the world (Bennett 2010). Western writings about earthworms date back to Aristotle, who suitably called them 'the Intestines of the Earth'. And, in the late 1800s, Darwin pointed out that the work of worms is vital for maintaining soil fertility in ecosystems such as grasslands, forests, and agriculture. As prolific eaters that work in symbiosis with bacteria, these vermicular invertebrates break down dead plant and animal matter in soils and continuously turn soil over to preserve its structure, aeration, drainage, and fertility (Edwards 2004). They also move large quantities of soil from the deeper strata to form a stone-free layer on the surface of the soil. Through this significant work, earthworms leave cracks and crevices in the soil to maintain aeration, drainage, and porosity. The burrows they create are coated with a mucous that forms a suitable environment for microbes. As earthworms digest the organic matter, the nutrients present in it are converted to a form that is more bioavailable to plants (Vergara 2012). In other words, through their work of consuming matter, fragmenting it, and mixing it with mineral particles in soil to create water-stable aggregates, earthworms are essential to making the organic surface on earth upon which all terrestrial life depends.

While earthworm taxonomy is still incomplete, taxonomists have identified more than 8,000 species of earthworms from approximately 800 genera which live all around the world, except in regions with extreme climates, such as deserts and ice-covered areas (Edwards 2004). Earthworms are also quite diverse in size: Edwards (2004, p. 4) notes that 'no other terrestrial invertebrate has such a wide range of sizes between the smallest and the largest individuals'. In British Columbia, scientists have identified 24 different groups of earthworms. Eighteen of these are introduced European species; only four (*Bimastos lawrenceae, Arctiostrotus perrieri, Arctiostrotus vancouverensis*, and *Toutellus oregonensis*) are native to the province (Klinkenberg 2017). Most of the native species live in the wet, temperate hemlock zone on the west coast of British Columbia (Marshall and Fender 2007). This is also the place where we conduct our multispecies ethnographies. Another species that abounds in the urban environs of the early childhood centres where we conduct our study is *Eisenia fetida*. This is the introduced composting worm, most commonly known as the red wiggler because of its reddish/ purple colour.

The children encounter different kinds of earthworms all the time. They are in the gardens they tend, the forest they visit weekly, the compost bin they care for, and even on sidewalks after the rain. The earthworms always attract the children, who are fascinated by their slimy bodies and wriggly movements and often try to pick them up with a stick or with their hands to get a closer look. As the adults around watch the children handle the worms, some of them struggle to contain their disgust.

For the most part, the children are oblivious to the ways in which the worms negatively affect the adults. They remain focused on the earthworms' responses to their touch as they quickly wriggle away down beneath the soil, or squirm on their hands, or droop from the end of a long stick. The children interpret these responses in many different ways. Sometimes they jump and giggle as they comment on how the worms tickle their hands. Other times they are determined to find the earthworms 'a house', or name them, or bring them to what they call 'safety'. Yet other times, they simply spend long moments enjoying the worms' movements in their hands as they softly share their own stories with them.

Without necessarily knowing anything about the impressive historical, scientific, or ecological significance of earthworms, these sticky beings still capture the children's attention and interest. Why are the children so curious about earthworms? How much of their curiosity is based upon the embodied alterity of these small, slimy creatures, and their desire to touch them? What possibilities reside in different forms of relating to earthworms, particular affect-based relations that require embodied encounter?

From experience, the children and the educators know where to find earthworms, and they visit these places frequently in search of the vermicular creatures. In the forest, the children like to frequent a shallow stream where there are always many native worms. They have named it Worm River. Wearing their colourful rubber rain boots, the children walk around in the water very carefully to ensure that they don't squash the worms. They also know that digging under the forest litter will yield many treasures, including earthworms. They love to unearth, inspect, and then put them back into the forest's rich soil. One of their favourite places to dig is in the rotten root system of a huge fallen cedar tree. When they dig into the crumbling roots, they can see how the tiny earthworms are gradually helping to decompose the enormous tree.

We wonder if the children are getting the sense that not all forms of power are determined by size. We know that the children are aware that their own bodies are so much bigger and heavier than even the fattest earthworms, because they take care not to tread on them. But the children's bodies, in turn, are dwarfed by the giant cedar trees. They too would be squashed if a tree fell on them. And yet, it is the same tiny earthworms that they can so easily squash who eat away at the massive fallen tree, turning it back into soil, powerfully transforming the very substance of this imposing forest. Perhaps their embodied encounters with worms as both squashable and forest reshaping are registering a visceral, affective appreciation of who is vulnerable to whom.

Each encounter with earthworms is unpredictable, with unintended consequences, dilemmas around responsibility, and also some elements of surprise. Questions about who gets to live, who gets to die, and how life is sustained emerge through these encounters.

Although they like to care for them, find new houses for them, or transport them to areas where they will not be stepped on, accidents happen. . . . One morning during one of our visits to the forest, a child was tending a wiggly worm in Worm River. He picked up the worm with a long stick to ensure that other children running around in the water wouldn't squash it. As it was being transported, the worm fell from the end of the stick. Determined to complete the task, the child struggled to pick it up again with the stick. This time, though, perhaps because of overzealous handling, the unlucky worm broke in two (see Figure 3.1).

Figure 3.1 A broken worm.

Author's (Veronica Pacini-Ketchabaw's) photograph

This unintended outcome was distressing for the child, particularly so because he was trying so hard to care for it by mitigating against its vulnerability to human feet. Sometime later, he was surprised and somewhat relieved to learn that an earthworm that is cut into two pieces can regenerate at least one half of itself, like some plants can (see Xiao *et al.* 2011). We wonder if this 'worm magic', as the child called it, might present new possibilities for thinking about earthworm vulnerability and resilience together? Might we think of the potential for regeneration, not just for the earthworm, but also for the children? Might it enable the children to reimagine life, precarity, and resilience anew? In other words, perhaps the 'magic' of earthworm regeneration might prompt ways of apprehending life and death beyond dualistic humanist conceptualisations, and in so doing, begin to shift our understanding of our own vulnerabilities relative to other beings.

The children have many unintended encounters with worms, including mortal ones. They often step on them, by accident, while walking on wet sidewalks – especially on rainy days – but are differently affected when they do. Some children move the injured worms to the grass. Others become upset when they see the squished worm beside their feet. What is interesting for us is that the children *notice* the worms, notice that they have an effect on the worms (for instance, when they poke them, pick them up, accidentally stepping on them), and also notice that the worms affected them (for instance, when they feel upset that they have hurt them or feel pleasure when the worms tickle their hands). What does the children's capacity to notice and be affected by the worms mean and do?

We think it has something to do with being drawn into relationship. We recognise that not all encounters are innocent and that some individual lives are more precarious than others. For instance, all large walking animals inadvertently step on and kill small creatures, whether they are aware of it or not. However, the fact that the children notice that they sometimes squash worms as they walk, and that this same noticing affects them, draws them into a relationship of responsibility for the life and death of others – the foundational awareness of all ethics. An ethics of responsibility comes from paying attention to our implication in the life and death of others, but it is not about aiming to be righteous, pure, and perfect. No being can exist without killing others. We are reminded of Shotwell's (2017, p. 8) writings about impurity as a mode of 'living ethically in compromised times'. She suggests that we follow 'impure' practices of paying attention, because such non-righteous modes of noticing ensure that we stay open to 'the field of possibility that might allow us to take better collective action against the destruction of the world in all its strange, delightful, impure frolic' (p. 8).

Picking up on the children's intent curiosity about the earthworms they encounter on the walks, the educators decide to extend their first-hand worm experiences. They arrange to purchase a compost bin and worms from the local compost education centre and to set it up in the childcare centre. This provides the children with the opportunity to explore the possibilities for multispecies collective action in very close quarters with thousands of red wiggler earthworms. The red wigglers' contribution to the collective action is to do the actual composting: to turn the food scraps into soil. But the children also contribute to the vermicomposting by taking good care of the worms.

There is a lot to do to ensure that the worms survive. The children help to shred newspapers for the worms' bedding, they collect dry leaves to mix with the food scraps, and they keep the bin moist. They learn what to feed and what not to feed the worms: raw fruit and vegetables but not meat or oily food; coffee grounds and egg shells but no dairy. They also learn how much to feed the worms: no more than two 750 grams containers full of food scraps per week.

The most labour-intensive part of the children's contributions to the vermi-composting process takes place each month. This is when they help to harvest the worm castings, which are then used to fertilise their garden. They dump the surface contents of the bin onto a big tarp. Then they work for hours, carefully removing any remaining worms, putting them back into the bin, and separating the castings into small piles.

In these very practical ways, the children care for the composting worms. They also come to understand what the worms need by observing how they respond to their care. For instance, if they put too much water into the bin, the worms cannot do their job, the food scraps start to rot, and the compost bin smells bad. If they put too much citrus in the food or do not give the worms enough food, the worms start to die. If there is not enough air in the bin or the bin is too cold or too hot, the worms also start to die. If they don't bury the food or if they overload the bin, the fruit flies get in and compete with the worms. At harvesting time, they also learn that the worms emit a foul smell when disturbed, and that it's easier to harvest outside because the worms retreat from the sunlight and dive back down into the bottom of the bin.

The significance of this is not that these children are learning to become more caring, and thus better humans, through tending to the worms. The significance lies in the fact that through taking part in the 'collective action' of vermicomposting, they have the opportunity to notice how composting worms are both vulnerable and responsive to a multitude of variables in their environment, and from this they learn how to adapt their own contribution to the collective project of making good compost. Moreover, as Maria Puig de la Bellacasa (2010, p. 152) writes, it is only in 'everyday practices at the level of ordinary life' that an ethics of inter-species care and obligation can grow. So perhaps it is through these daily routine practices of feeding and looking after the worms (who in turn respond by making fertilised soil for the garden) that the children consolidate an ethics of 'caring obligation' towards the worms. The children's obligation is to care for the worms through feeding them and maintaining a healthy bin environment. In return, the worms care for them through making fertile soil to grow more food.

Thinking with children and ants

Ants might be small, even smaller than earthworms, but they render big species like us humans quite insignificant in other scalar terms. They predate us by at least 100 million years and outnumber us by around 1.5 million to one (Ward 2010).

They are the most numerous and species-diverse of all social insects. Apart from their primordial and ubiquitous presence in all parts of the world (except for the polar regions), they are probably best known for their extraordinary social organisation. Ants' high-level sociality is evident in their capacity for industrious collective work, which they undertake within complex systems of labour division, communication, and cooperation. It is because of their capacity to act collectively that ants can effectively colonise and modify environments. One ant alone cannot make a huge impact on the world, but working collectively, ants become a powerful engine room of earth movers and shapers.

There are gravelly ant nests all through the dry sclerophyll bushlands where we walk each week. The children never lose their fascination for them (see Figure 3.2). On every walk, they stop, bend over, and stare intently into the miniature worlds of tiny pebbles and small holes that cover the surface of the underground ant homes. They are looking to spot the scurrying inhabitants as they dart in, out, and about. In the cooler months, when the ants are more docile, the children squat down on the ground to inspect the nests at close range. They study the tiny component parts of this micro-landscape in minute detail, but always with an eye for the action. They often call out when they spot ants carrying dead insects and seeds into the holes. When the ants are sluggish, the children push small twigs down the holes to see if they can rouse them. They know that there are thousands of ants down there somewhere.

What is it about the small details of this miniature ant world that repeatedly grabs and holds the children's attention? Does all this close looking give them an inkling that there are complex and mysteriously unknown worlds beyond their own? Are they somehow registering that worlds exist within worlds, and, if so, do they have a sense that these nested worlds are connected?

In hot, dry environments, such as those that predominate in Australia, ants do the work of worms. They are the subterranean 'ecosystem engineers' that keep the soil structure healthy and porous by digging aerating tunnels and allowing precious rain to penetrate the ground (Evans *et al*. 2011). Like worms, they are also extraordinary composters. Worker ants clean the surrounding surface environment and increase the nutrient base of the soil by depositing plant and animal waste in their tunnels for food storage. They are extremely efficient at doing this regenerative work. Each ant can carry more than 20 times its body weight (Ward 2010). Many ants also have close and mutually beneficial relationships with plants, usually involving the exchange of protection for nutrition. The most famous example of this is the well-documented symbiotic relationship between seed-gathering ants and acacia plants. The ants increase the dispersal and germination rate of the acacia seeds by carrying them down into their nests and effectively 'planting' them in the nutrient-rich soil. In return, the ants get to feast on the acacia seeds' sweet outer top casing, called the elaisome, before the seeds germinate (Forest and Madden 2011, pp. 2–3).

As the weather warms up, so do the ants. They come out and about, doing their business. They get faster and feistier. The children quickly learn that when they stomp on the ground, the ants will come running out. As the agitated ants start to

Figure 3.2 The children are fascinated with the ants.

Author's (Affrica Taylor's) photograph

swarm over the top of the nest, protecting their territory, the children back off to a respectful distance. But they still can't resist prodding the nests with sticks, hoping for a rise, or perhaps the thrill of seeing a line of ants marching up the stick towards their hands.

It is interesting to note the ways in which both children and ants test their agency in this encounter and how this agency is determined, not only by size but also by speed and volume. By partaking in this provocative dance of relating, of threatening and protecting, of advance and retreat, what do the children learn about the relative power and effect that large and small creatures can exert upon each other?

Sometimes, ants run up the children's legs or into their clothing and bite them. This is usually a form of retaliation against the children's provocations, or when the children simply don't notice that they're standing in the middle of a swarming ant nest. There have been some highly charged moments when frenzied ants scurry and bite and panicked children scream and squash. These fight and flight, life and death moments are marked by the rush of alarm pheromones and adrenalin and by the smell of formic acid. But the children who calmly observe the ants rarely get bitten. One child is waiting patiently for an ant to walk onto her hand, so she can take an even closer look at it. Her perseverance pays off. Both ant and child seem unfazed by their intimate encounter.

Both ants and children face a risk and pose a threat to each other in these embodied exchanges, albeit unevenly. Both register their mutual vulnerabilities. While the defensively swarming ants clearly recognise the children as a potential threat to their nests, the children are well aware that ants can bite them. They also know they can easily kill ants. This understanding presents an ethical dilemma for the children: to kill or not to kill the ants, before or if they bite them? It seems the children are on a number of different routes towards responding to the risks and vulnerabilities they share when they bodily engage with these small creatures. Their actions portend different kinds of learnings. The children who goad ants might learn that there are consequences to their actions and that even small creatures can become formidable foes. Those children whose feet inadvertently get in the way might learn the consequences of not paying attention to the lifeworlds of smaller creatures. Those who carefully seek intimacy with the ants might learn about the precarity of life through (literally) holding the responsibility for another life and, at the same time, through risking making themselves vulnerable to another species.

Conclusion: a common worlds ethics of multispecies vulnerability

When responding to the Anthropocene in this chapter, we have been very conscious of scale – the scale of the problems of the Anthropocene, the scale of children's immediate lifeworlds, the scale of worms' and ants' lifeworlds, and the scale of our own efforts to respond. We are well aware that we are engaging as very small players, along with very tiny animals, on a very small stage, with very small interspecies events, in the face of gargantuan planetary challenges. Our primary scalar strategy has been, and still is, to situate our research within the geo-historical specificities of the immediate commonplace worlds that children actually inhabit along with other species and to gradually unravel and trace the entangled threads that knit this particular common world into countless other worlds – and, indeed, into the global commons.

Our modest efforts have been buoyed by Donna Haraway's (2013) reminder that only 'partial recuperations' are possible and that responding to the Anthropocene is as much about paying close attention to everyday small things, contingent partialities, and messy relationalities as it is about the geo-sublime proportions of carbon measurements, global warming, and melting ice caps (Instone 2015). We have also been invigorated by what we have learnt from ants and worms. They have reminded us that, beyond the radar of the spectacular 'big action' we partake in on the surface of the planet, a proliferation of tiny earth movers, industriously working underground, taking collective action, are literally reshaping, shifting, and making the earth.

As researchers, our abilities to notice and appreciate the complex minutia of interspecies reciprocities and co-shapings have been sharpened and refined by witnessing what happens when the bodies of muddy children and slippery worms touch each other in the wet forests of British Columbia, and when dusty ants and children congregate in the dry bushlands of the Australian Capital Territory. These seemingly minor, trivial, and yet also mortally intense earthy encounters have shown us that learning how to respond and respect in multispecies common worlds can only happen in those embodied (and often fraught) moments when humans and animals actually meet and notice each other.

Our witnessing of such real-life encounters amongst children, ants, and worms and our readings of Myra Hird's (2010) microontologies have, in combination, helped us to conceive of ways in which we might bring very small animals into the realm of ethics within these common worlds. We may have not been able to completely resist the temptation to default to a self-referential humanist ethics of largesse and care for these little creatures, but at least the not-always-gorgeous encounters with faceless ants and worms, and an apprehension of their incredible collective potency, gave us some sense of the possibility for tapping into an ethics of our own vulnerability to other lifeforms on earth.

All of the aforementioned have helped us glimpse how we might remake worlds with other species in the Anthropocene and work towards a common-world-attuned ethics of mutual vulnerability. The point we are making here is that, despite the human predilection to reiterate human exceptionalism, including within many epic and heroic narrations of the Anthropocene, the fact is that our human lives are totally dependent on the lives of other, much smaller, often overlooked, and sometimes invisible creatures, such as worms and ants. Their work in composting the earth to make it viable for other lifeforms not only predates our own relatively short, if spectacular, human life on earth, but will most likely postdate us as well. Learning with other species would not exclusively follow human plans, wishes, and desires. Everything becomes relevant in fostering and researching child–animal relations in the Anthropocene, human, and otherwise. As Tsing (2013) writes, the challenge is bringing 'more-than-human socialities' into our understandings of pedagogies and research.

But we have to admit that this is no easy task, and there have been limits to our work. These glimpses are hard to sustain. Taking the encounters amongst children, earthworms, and ants as serious moments of multispecies mutual vulnerability has required us to continually work at decentring the human in our own thinking and practice. While it has been quite easy to recognise the agency of the ants and worms within the encounters (particularly those biting ants), we have found it nigh impossible not to position the children as the central actors and difficult, at times, to resist the temptation to search for evidence of an ethics based in human-centric care for the other. Despite these limitations, we continue to argue for an ethics of mutual vulnerability in the Anthropocene that recognises that learning through encounters with other species is not always harmonious and pleasant, is not always equal, and does not offer us 'moral certitudes or simple escape routes' from the mess we are in (Haraway 2011, p. 115). Perhaps the most we can claim is that these small interspecies pedagogical encounters are incremental, if fleeting, moments of 'multispecies achievements' (Haraway 2013), 'where multispecies players enmeshed in partial and flawed translations across difference redo ways of living and dying' and of re-worlding damaged worlds together.

References

Bennett, J., 2010. *Vibrant matter: A political ecology of things*. Durham: Duke University Press.

Edwards, C., 2004. The importance of earthworms as key representatives of the soil fauna. *In*: C. Edwards, ed. *Earthworm ecology*. New York: CRC Press, 3–11.

Evans, T.A., Dawes, T.Z., Ward, P.R. and Lo, N., 2011. Ants and termites increase crop yield in a dry climate. *Nature Communications*, 2, 26. doi:10.1038/ncomms1257.

Forest, F. and Madden, D., 2011. *Elaiosomes and seed dispersal by ants*. DNA to Darwin Case Study. Available from: www.dnadarwin.org/casestudies/2/FILES/AntsSG2.0.pdf [Accessed 28 February 2018].

Haraway, D.J., 2011. Speculative fabulations for technoculture's generations: Taking care of unexpected country. *Australian Humanities Review*, 50, 95–118.

———., 2013. *Distinguished lecture*. Arizona State University, Institute for Humanities Research [video]. Available from: https://ihr.asu.edu/node/540 [Accessed 28 February 2018].

Haraway, D.J., Ishikawa, N., Gilbert, S., Olwig, K.R., Tsing, A.L. and Bubandt, N., 2015. Anthropologists are talking – About the Anthropocene, *Ethnos: Journal of Anthropology*, 81(3), 535–564.

Hird, M.J., 2010. Meeting with the microcosmos. *Environment and Planning D: Society and Space*, 28, 36–39.

———., 2012. Animal, all too animal: Toward an ethic of vulnerability. *In*: A. Gross and A. Vallely, eds. *Animal others and the human imagination*. New York: Columbia University Press, 331–348.

———., 2013a. *On preparations: Learning and teaching materiality*. Paper presented at Developing Feminist Post-Constructivist Qualitative Research Methodologies in the Educational Sciences Conference, Stockholm University, June 2013.

———., 2013b. Waste, landfills, and an environmental ethics of vulnerability. *Ethics and the Environment*, 18(1), 105–124.

Instone, L., 2015. Risking attachment in the Anthropocene. *In*: K. Gibson, R. Fincher and D.B. Rose, eds. *Manifesto for living in the Anthropocene*. New York: Punctum Books, 29–36.

Kind, S., 2013. Lively entanglements: The doings, movements, and enactments of photography. *Global Studies of Childhood*, 3(4), 427–441.

Kirksey, S.E. and Helmreich, S., 2010. The emergence of multispecies ethnographies. *Cultural Anthropology*, 25(4), 545–576.

Klinkenberg, B., ed., 2017. *Earthworms*. E-Fauna BC: Electronic Atlas of the Fauna of British Columbia. Lab for Advanced Spatial Analysis, Department of Geography, University of British Columbia, Vancouver. Available from: http://ibis.geog.ubc.ca/biodiversity/efauna/ [Accessed 28 February 2018].

Latour, B., 2004. How to talk about the body: The normative dimensions of science studies. *Body and Society*, 10(2–3), 205–229.

Lorimer, J., 2010. Moving image methodologies for more-than-human geographies. *Cultural Geographies*, 17(2), 237–258.

Marshall, V.G. and Fender, W.M., 2007. Native and introduced earthworms (Oligochaeta) of British Columbia, Canada. *Megadrilogica*, 11(4), 29–52.

Ogden, A., Hall, B. and Tanita, K., 2013. Animals, plants, people, and things: A review of multispecies ethnography. *Environment and Society: Advances in Research*, 4, 5–24.

Puig de la Bellacasa, M., 2010. Ethical doings in naturecultures. *Ethics, Place, and Environment: A Journal of Philosophy and Geography*, 13(2), 151–169.

Schwartz, J., 2012. *Worm work: Recasting romanticism*. Minneapolis: University of Minnesota Press.

Shotwell, A., 2017. *Against purity: Living ethically in compromised times*. Minneapolis: University of Minnesota Press.

Stengers, I., 2016. *In catastrophic times: Resisting the coming barbarism*. London: Open University Press.

Tsing, A.L., 2013. More than human sociality: A call for critical description. *In*: K. Hastrup, ed. *Anthropology and nature*. New York: Routledge, 27–42.

Vergara, S.E., 2012. Worms. *In*: C. Zimring and W. Rathje, eds. *Encyclopedia of consumption and waste: The social science of garbage*. Thousand Oaks: SAGE, 1008–1010.

Ward, P.S., 2010. Taxonomy, phylogenetics, and evolution. *In*: L. Lach, C. Parr and K.L. Abbott, eds. *Ant ecology*. Oxford: Oxford University Press, 3–17.

Whatmore, S., 2013. Earthly powers and affective environments: An ontological politics of flood risk. *Theory, Culture, and Society*, 30(7/8), 33–50.

Xiao, N., Ge, F. and Edwards, C., 2011. The regeneration capacity of an earthworm, *Eisenia fetida*, in relation to the site of amputation along the body. *Acta Ecologica Sinica*, 31, 197–204.

4 Children, bilbies, and spirit bears
A decolonising ethics of ecological reconciliation

> The newcomers do not understand the land the way the original people do. So we must reach the children. They will understand . . . [we are] asking them to help us stop the invaders and give new life to the land.
>
> (Bright 2012/1993, p. 3)

Animal characters in settler colonial children's literature

To supplement the staple animal characters from the 'mother country', children's literature and popular culture in ex-British settler colonial countries have produced a steady stream of native animal protagonists. Since the mid-20th century, the amusing antics of 'homegrown' native animal characters – such as Yogi Bear, Wile E. Coyote and the Road Runner in North America, and the Muddle-Headed Wombat and Skippy the Bush Kangaroo in Australia – have offered children new modes of identification with their own settler cultures. They have helped to secure these children's affections for and affinity with the non-European animals and natural environments in which they live, and hence with the settler colonial nation. However, even though these native animal characters were created to represent the distinctive 'natures' of their respective settler nations, they commonly retained some of the core generic traits of the imperial prototypes. As with Pooh Bear and Peter Rabbit before them, many feature anthropomorphised native animal protagonists who wear clothes, speak English, are caricatures of human foibles, and are embroiled in human-mimicking story lines and dramas designed primarily to entertain children. In recent times, though, there has been a discernible shift in the mode of representation of native animal characters.

The new trend in settler colonial children's literature and popular culture is to produce more ecologically motivated native animal narratives, which not only entertain children but also inform, educate, and interpolate them as native animal allies. Recast within the vernaculars of wilderness, eco-nationalist, and conservation discourses, the native animal characters of this new eco-genre of children's animal fiction are often portrayed as living in environments damaged by colonisation and/or precariously surviving in increasingly threatened natural habitats. The dramas that unfold typically revolve around the vulnerability of the native animal characters and their plight to survive the ravages of colonisation. Concomitantly, their predominantly non-indigenous young readers and audiences are interpolated as their protectors and/or saviours, as the new breed of environmental stewards. The alliance that is fostered between the settler children and indigenous wildlife is typified in the opening quotation where, in the assumed absence of indigenous custodians, children are identified as those most likely to understand the plight of endangered native animals and handed the responsibility to 'stop the invaders and give new life to the land' (Bright 2012/1993, p. 3).

In this chapter, we turn a spotlight on the ways in which non-indigenous children, along with indigenous and non-indigenous animals, are co-implicated within contemporary settler colonial environmentalist tropes of invasion, destruction, and salvation. To unpack the complex interactions of these intersecting tropes, we examine children's and young people's narrative texts that feature two endangered native species – bilbies in Australia and spirit bears in Canada. We pay particular attention to the ways in which the narratives align endangered native animals, indigenous people's knowledges and beliefs, non-indigenous children as the threatened or rescuing 'goodies', and feral non-indigenous animals and environmentally exploitative (non-indigenous) adults as the invasive or destructive 'baddies'.

We start by tracing how efforts to save the bilby emerged from an Australian environmental movement aimed at eradicating feral rabbits and then turn to examine a range of children's bilby books associated with this movement. These include picture books aimed at very young children – *The Bilbies' First Easter* (Sibley 1994) and *The Smallest Bilby* series (Hilton and Whatley 2006, 2009, 2014) and books for early readers – *Burra Nimu, the Easter Bilby: A Story for Australian Children* (Bright 2012/1993) and the *Bilby's Ring* trilogy (Kessing 2015a, 2015b, 2015c). We reflect upon the fraught settler colonial politics of positioning rabbits as the primary invasive threat to the bilby and enlisting non-indigenous children as the bilby's primary protectors and saviours.

We then move on to explore the distinctive Canadian style of wilderness conservation that has been deployed to save the British Columbian Kermode or 'spirit' bear. We look at *Spirit Bear: The Simon Jackson Story* (2005), a film associated with the Spirit Bear Youth Coalition campaign to save the spirit bear. The narrative features of the film and the publicity campaign surrounding it allow us to reflect upon the ways in which indigenous knowledge and spirituality is appropriated and surpassed within contemporary ecological narratives and how the responsibility of stewardship for threatened ecological systems and species is placed firmly in the hands of non-indigenous youth.

We situate this new breed of endangered native animal texts that enlist children as environmental stewards and saviours within a broader analysis of the cultural politics of settler environmental stewardship and against the backdrop of the current anthropogenic sixth mass species extinction event that is so closely associated with colonisation, which we discussed in the introduction. Drawing on recent feminist critiques of the human-centric solipsism of stewardship discourses (Stengers 2015; Tsing 2015; Haraway *et al.* 2015; Haraway 2016) we question both the premises and the implications of these endangered species children's texts. We conclude the chapter by considering how we might interrupt and augment these well-meaning but politically fraught endangered species stewardship narratives with a decolonising ethics of ecological reconciliation (Rose 2004, 2011).

Figuring a new breed of endangered native animal characters

Ecological awareness is changing the landscape of Australian and Canadian children's animal literature and popular culture. A new breed of endangered native animal characters is being figured within an emergent body of children's eco-literature and popular culture. In addition, the targeted child/youth readers and audiences of these eco-texts are increasingly positioned as the protectors and saviours of endangered native animals, often quite explicitly within the narratives themselves. In Australia, there is a growing body of stories featuring endangered bilby characters that is directly appealing to settler Australian children's environmental sensibilities and patriotic loyalties. In Canada, the British Columbian spirit bear is superseding the generic, anthropomorphised bear as a key protagonist in ecologically attuned cultural texts aimed at educating settler Canadian children. So who are the real bilbies and spirit bears that inspire this new fictional breed of endangered native animal characters?

The great bilby, or *macrotis lagotis*, is a small, grey, burrowing marsupial. Because of its large, rabbit-like ears, its hopping gait, and the fact that it digs its own underground home, the bilby is also known by the common name of 'rabbit bandicoot' (Flannery *et al.* 1990). Before colonisation, bilbies lived across 70 percent of the Australian continent. Today, they are confined to 15 percent and live in isolated colonies within the arid sandy desert regions, making them particularly vulnerable. The greater bilby's close relative, the lesser bilby, is already extinct, and its own numbers are in a state of continual decline. There are fewer than 10,000 greater bilbies left (Threatened Species Network n.d.).

The Kermode bear (*Ursus americanus kermodei*) is a rare North American black bear which is only found on the northwest coast of British Columbia and in southwest Alaska, predominantly in British Columbia's Great Bear Rainforest. It is also known as the spirit bear (McAllister and Read 2010). Spirit bears appear to be a 'a walking contradiction' because they are *white* black bears (Barcott 2011). Their white coats are the result of a recessive mutation in the MC1R gene within the black bear population of the coastal rainforest and island regions of the province (Ritland *et al.* 2001). Both parents need to carry the recessive gene in order to produce a white offspring, and even then, the chances are very low. As a result, there are fewer than 400 spirit bears in the whole Great Bear Rainforest, making them one of the rarest animals in the world (Barcott 2011).

Despite their obvious species and ecological distinctiveness, bilbies and spirit bears have a lot in common. They share a combination of attributes that has brought them popular, if not iconic, status in their respective countries. They are both extremely rare native animals that live in wild and remote areas and are therefore hardly ever seen in their natural environments. Most adults and children learn about bilbies and spirit bears from cultural texts. Neither of these species constitutes any form of threat to humans and both have charismatic physical appeal. They are therefore very easy to love (at a distance) and to be regarded as highly deserving of protection – not the least because their habitats have been or are about to be damaged by settler activities.

In Australia, settler clearing and the spread of introduced feral animals – namely European wild rabbits, foxes, and cats – have already decimated the bilby population. Rabbits were initially introduced by gentrified British settlers in the 19th century for hunting purposes, but very quickly became feral and spread across the continent in plague proportions, eating everything in sight. In Australia, wild European rabbits are identified as one of the key destructive 'invasive species'. And, as we will show in this chapter, bilbies are seen as the key native species victims of invasive European rabbits.

The situation in Canada is somewhat different, as spirit bears have always been rare and were regarded by First Nations people as special well before colonisation. The Tsimshian peoples call the spirit bear Moksgm'ol. They believe that spirit bears were created by Raven, the trickster creator of the rainforest, who turned every tenth black bear white as a reminder of the last ice age. Non-indigenous Canadians were largely unaware of the spirit bears' existence until the early 1900s (McAllister and Read 2010) when logging started in the Great Bear Rainforest region. The rare spirit bears' survival was threatened from this moment, along with the ecological integrity of the entire forest environment. Further destruction of this fragile environment was anticipated when construction of the Northern Gateway Pipeline to carry oil from Alberta to west coast ports was planned to be routed straight through the rainforests (Howlet *et al.* 2009). Because it is so rare and special, the spirit bear has become a symbol of the vulnerability of pristine Canadian wilderness to the destructive forces of settler logging and fossil fuel extraction.

Saving bilbies

Since early settler colonial times, the bilby's history and fate has been materially and semiotically entangled with that of the wild European rabbit. In the last 30 years or so, this already entangled interspecies relationship has been further complicated by an idiosyncratic mix of eco-nationalist conservationist interventions, commercial interests, and cultural productions that have specifically targeted Australian children.

It all started in 1968, when publicity about a story called 'Billy, the Aussie Easter Bilby', written by a 9-year-old Queensland schoolgirl, Rose-Marie Dusting, captured national attention and spurred interest in and concern about the plight of the endangered bilby. The idea of the bilby replacing the rabbit as Australia's own Easter animal culminated in the launch of the Easter bilby campaign in 1991, spearheaded by an environmental lobby group that was subsequently named the Foundation for Rabbit Free Australia. It was not long before Australian chocolate manufacturers tapped into the potential new markets opened up by the Easter bilby campaign, and the chocolate Easter bilby was born (Haigh's Chocolates n.d.). More than three decades on, in the run-up to Easter, Australian supermarkets and chocolate shops still offer chocolate bilbies as a popular alternative to the standard chocolate eggs and bunnies. As the respective official sponsors of the Rabbit Free Australia (2017) and the Save the Bilby Fund (2017), Haigh's Chocolates and Darrell Lea Chocolates make it publicly known that they donate a proportion of their profits from chocolate Easter bilby sales to help 'save' the bilby. Their contributions help to fund projects aimed at protecting bilbies, such as rabbit virus eradication research, predator-proof fencing, captive breeding, and the tagging and monitoring of bilbies (Zielinski 2012).

Coordinated via its website, the Queensland-based Save the Bilby Fund (2017) conducts a wide variety of public awareness events and fund-raising activities. These include hosting a recent 'National Bilby Summit' to coordinate plans to 'save the bilby from extinction', organising an annual 'School's Bilby Day' and 'Go Green for Bilby' events, running a 'Bilby Buddy' sponsorship program to help support bilbies in captive breeding programs, and selling bilby merchandise. As well as their alliance with Darrell Lea Chocolates, they also partner with commercial zoos, wildlife parks, eco-tourist ventures, and eco-product producers.

In addition to its ultimate goal of eradicating European wild rabbits, the conservationist group Rabbit Free Australia (RFA) is on the associated mission of saving bilbies. Flagging an overtly eco-nationalist agenda, its website runs with the header slogan 'Bilbies not bunnies. Reclaiming the Australian environment'. Its stated vision is for 'Australian landscapes that are free of their most notorious pest – the European wild rabbit', its emblem is a bilby, and its acknowledged 'major sponsor' is Haigh's Chocolates. The interrelationships between rabbits and bilbies – furry, chocolate, and fictional – are elaborated in this way:

> The Easter Bilby champions the cause of Rabbit Free Australia – fighting back against rabbits and reclaiming a place in the Australian environment. RFA is the registered holder of the trademark 'Easter Bilby', and is pleased to work with chocolate manufacturers and story-tellers using Easter Bilby to promote the rabbit control cause and a theme of **'bilbies not bunnies'**.
> (Rabbit Free Australia 2017)

Children are clearly situated as the key audience and consumers within this peculiarly settler-Australian, bilby-rabbit-Easter-eco-nationalist-literary-confectionary assemblage. Australian children's story-tellers have played a significant role in cultivating the Easter bilby movement and in articulating and disseminating the theme of 'bilbies not bunnies' – whether as self-publishing (Bright 1993/2012; Garnett and Kessing 2006/1994; Kessing 2015a, 2015b, 2015c) or commercially published authors and illustrators (Sibley 1994; Hilton and Whatley 2006, 2009, 2014).

One the earliest publications to bring the figure of the bilby into Australian children's literature, and to associate it with Easter, was Irene Sibley's (1994) picture book *The Bilbies' First Easter*. The bilby characters in Sibley's narrative are shy but magical creatures. As rain falls at the end of a long and difficult drought, they bring hope and joy back to the lives of an embattled white-settler family living on a remote outback farm. They do this by using the colours of the rainbow to paint bright and cheerful eggs for the family's Easter celebrations. This pioneering bilby story is stitched into the traditional Australian settler trope of survival in the Australian bush. Told from the perspective of the main child character, it recounts the challenges and rewards of settler life in Australia's harsh but regenerative (and beautiful) outback nature. While Sibley's title and aspects of her story line have clearly drawn inspiration from the Easter bilby campaign, and carries nationalist overtones, her bilby characters do not convey any explicit conservation messages. They primarily serve a familiarising function, drawing the young readers' attention to the existence of these relatively unknown (at the time) 'rabbit-like' Australian animals and portraying them as special, rare, and charismatic creatures, but not explicitly endangered.

Nette Hilton and Bruce Whatley's (2006, 2009, 2014) *Smallest Bilby* picture book series also follows the Easter bilby theme, and sets out to foster young Australian children's allegiance to native animals through creating ample visual and thematic opportunities for feel-good positive affective identification. The first book, *The Smallest Bilby and the Midnight Star*, sets the scene for this positive identification by linking its young readers' sense of smallness, vulnerability, and needs for love and belonging with those of Billy Bilby – the series protagonist, who is affectionately described as 'the smallest bilby of all'.

The association of bilbies with rabbits and Easter only begins in the second book, *The Smallest Bilby and the Easter Games* (2009), when the Easter bunnies decide they need a rest from delivering eggs and hold the 'Easter games' to determine which Australian animal will take on their job. Through a process of elimination, it is decided that Billy Bilby would be the perfect choice. Not only does he physically resemble a rabbit, but this highly empathic small bilby also understands that the significance of the job of delivering eggs to children is to 'make them feel special'.

The Smallest Bilby and the Easter Tale (2014) is the third book in the series. It revolves around the misadventures and challenges faced by Billy Bilby and his team of small bilbies as they undertake the task of Easter egg delivery under the watchful eye of the 'biggest Easter rabbit'. Hilton and Whatley use some fairly conventional, sentimental, and anthropomorphising literary traditions to secure children's affection for and identification with the bilby. In particular, Billy Bilby's essentially human qualities of kindness, gentleness, empathy, braveness, and humility at times override the central theme of Australian belonging. His character primarily serves as a good citizen role model for Australian children (albeit it a specifically *indigenous* good citizen), but it does little to educate them about Australian native animal ecologies.

Jenni Bright's *Burra Nimu, the Easter Bilby* is of a very different ilk. Also inspired by the original Easter bilby campaign and written in the early 1990s, it was revised in 2012 and self-published on Bright's 'Burra Nimu, the Easter Bilby' website (n.d.) with the additional subtitle *An Australian Children's*

Story. In keeping with RFA's eco-nationalist political mission to deploy the Easter bilby figure to 'fight back against the rabbit' in order to 'reclaim . . . the Australian environment', Bright's narrative is uncompromising, polemic, and didactic. Her bilby characters have indigenous, not European names, and the main character, Burra, takes the persona of a lead activist. In order to mobilise all of the native animals, he urges, 'We need to stand together, to fight as one if the land is to survive – if we are to survive – we must *do* something' (Bright 2012/1993, p. 1).

The narrative is overtly constructed around the settler eco-nationalist tropes of invasion, destruction, and salvation and the dichotomies of 'good' versus 'evil' that support them. The 'rabbit army' is quickly identified as the evil colonial invading force that has caused all of the environmental destruction, and Burra is the crusading environmental saviour. However, he also knows that he and his gang of native animals will need help to save the environment from the rabbits. So he makes a plan to paint eggs with all the beautiful colours and motifs of the natural environment to take to the children and enlist their support. In return for this gift, the children will help the native animals to save their lands and ensure the survival of the endangered native species. The children will be allies to the native animals and ultimately their saviours.

Although the story ends as soon as the native animals reach the children, it is clear from the narrative that they are the future environmental stewards and saviours: '"We've found the children!" Everyone gave a huge cheer. . . . There was hope now. The future lay ahead!' (Bright 2012/1993, p. 17). The majority of the narrative recounts the treacherous journey to reach the children, which is full of negative affect. The bilbies' ongoing military-style battles with the 'rabbit army' underscore the rabbits' inherent malice and work to extinguish any emotional attachments that young readers may have had to the bunny rabbit. The rabbits are pure evil. They do nothing but invade, destroy, ruin, and scream murderous things like 'Kill the bilbies! Capture the eggs! Stop them from reaching the children!' (Bright 2012/1993, p. 5). It is also noted that they 'kill each other's young to win more space in the burrow' (Bright 2012/1993, p. 5), driving home the notion that unlike the caring native animals, these invading alien rabbits are totally bereft of any kind of moral codes.

Similarly uncompromising in its vilification of non-native colonising feral animals, Kaye Kessing's (2015a, 2015b, 2015c) *Bilby's Ring* trilogy also tells a story of a brave bilby leading a band of native animals on an epic journey: *Out of the Spinifex* (Kessing 2015a), *Across a Great Wide Land* (Kessing 2015b), and *Into the Bowels of the Biggest City* (Kessing 2015c). Bilby's quest is 'to find humans with the Kindness inside, those who will listen and understand' (Kessing 2015c, p. 78). These are humans who can learn to

> care more for bilbies instead of rabbits; chuditch instead of cats. We must
> steal the kindness and caring in their hearts away from the invaders . . . so
> humans will care more for us and save us from the invaders.
>
> (Kessing 2015c, p. 79)

Along the way, Bilby and his band of 'natives' are constantly battling with and escaping from the marauding invaders, or 'ferals' (rabbits, foxes, and cats). At the end of the journey, when they reach the 'biggest city', they are rescued from the urban 'ferals' by Tinny, an Aboriginal boy, and Nessa, a white-settler girl. These good young humans also help them to find a way to get into a television station and to broadcast their message to the human world. Kessing's narrative is structured by the same eco-nationalist tropes of invasion, destruction, and salvation that dominate the *Burra Nimu* (Bright 2012/1993) story – minus the Easter egg theme. Aimed at older child readers, the series is additionally fortified by its intertextual nods to Tolkien's epic adventure quest trilogy *Lord of the Rings* and contains relentless graphic, violent encounters between the brave and heroic 'natives' and the evil 'ferals'.

Both Kessing's and Bright's eco-nationalist narratives rely heavily upon what Kate Wright (2014, 2017) refers to as 'redemptive violence'. As Wright explains, redemptive violence serves the function of atoning for past settler colonial mistakes (in this case the settlers' introduction of European wild rabbits), while keeping the spotlight firmly focused on the present (in this case on the rabbits, not the settlers who introduced them). By shifting the spotlight from past to present violence, from settler invasion to feral animal invasion, settler Australians not only escape taking responsibility for the cascading ecological effects of introducing rabbits in the first place, but simultaneously recast themselves and their progeny as the redemptive heroes – the ones who will save good 'native' Australia from the evil 'invasive' rabbits. The redemptive violence that is enacted in these narratives effectively performs a form of historical disavowal. Fortified by the force of their negative affect, horrified child readers are entreated to disidentify with the brutality of the invaders and to identify wholeheartedly with the embattled and endangered indigenes.

The battle lines that are drawn within these violent animal narratives delineate an uncompromising binary moral order of 'essentially good native' and 'inherently evil feral invader'. These moral binaries underpin the invasion/destruction/salvation tropes that pervade *Burra Nimu* and *Bilby's Ring*. Although the bilbies and their native allies are far from passive victims, it is clearly indicated that they cannot ultimately save themselves. Hope for their salvation inevitably lies in the hands of good humans – either the (assumed to be inherently good) children (Bright 2012/1993) or humans with 'kindness in their hearts' (Kessing 2015a, 2015b, 2015c). Both *Burra Nimu* and the *Bilby's Ring* series declare that because the indigenous custodians can no longer look after the land (good), settler Australians (particularly children) must now assume the role of native species allies/ protectors/saviours and environmental stewards.

We find the 'black and white' premises of these invasion/destruction/salvation narrative tropes to be fraught and dubious. When predominantly addressing a young *settler* readership, they are additionally paradoxical and demand a challenging set of subject repositionings and loyalty realignments. The key strategy to affecting such moves is their eco-nationalist appeal. In order to prove that they are loyal, patriotic Australians, settler children must relinquish the bunny for the bilby – choose native Australian over European. Moreover, in order to become truly naturalised, to become 'authentically' Australian, settler children must learn how to step into the shoes of indigenous custodians and become the next generation of Australian stewards.

Saving spirit bears

> *February 3, 2016. Today I awoke knowing that a rare, remarkable and ecologically important bear will forever fish for salmon, sleep in the hollows of ancient trees and walk through the mist shrouded forests it has known since time began. For the first time in two decades, it can be said with confidence that the spirit bear is not just safe, but saved.*
>
> (Jackson 2016)

Evoking the sacrosanctity of the North American transcendentalist wilderness trope, Simon Jackson wrote this victory blog entry, declaring that 'the spirit bear is not just safe, but saved' to mark the success of an extraordinary youth campaign to save the spirit bear from extinction. He also wrote it to signal the end of the Spirit Bear Youth Coalition that spearheaded this campaign – a coalition that he founded in the 1990s and subsequently led for two decades. Originally a Canadian youth coalition, it grew dramatically over this time, achieving international acclaim. According to its own claims, it became 'the world's largest youth-led environmental movement with a global network of more than 6 million in over 70 countries' (Spirit Bear Youth Coalition n.d.).

The romantic sublime of North American wilderness discourses, evident in Jackson's blog (noted earlier), is a feature of the texts associated with the campaign to save the spirit bear and its pristine temperate rainforests. It can be seen, for instance, in the children's picture book *Spirit Bear* (Harrington and Arnott 2013), published by Eco Books 4 Kids (2013), a Canadian publisher with a stated educational and conservationist agenda. The book is accompanied by a teachers' guide and designed to introduce Canadian children to the plight of the spirit bear. The book works to plug its young readers into a sense of nostalgic identification with a rare and endangered animal and habitat that hardly anyone has ever seen or actually knows, by appealing to the sentiments of the settler colonial Canadian wilderness imaginary. Heise (2016) argues that narratives about threatened species are compelling because of the perception that 'part of national identity and culture itself seems to be lost along with the disappearance of a nonhuman species' (Heise 2016, p. 49). As she further explains, conservationist and stewardship tropes become a way of attending to

> worries about the future of nature, on the one hand, and on the other, hopes that a part of one's national identity and culture might be preserved, revived, or changed for the better if an endangered species could be allowed to survive.
>
> (Heise 2016, p. 49)

Apart from their distinctive appeal to the settler North American pristine wilderness imaginary, the key Canadian 'save the spirit bear' narratives share many features in common with the Australian 'save the bilby' ones. Like the Australian texts, they are predominantly targeted towards young audiences/readers, whom they explicitly identify as the bears' saviours and stewards. They deploy the same basic narrative tropes of (settler) invasion, destruction, and salvation, and are structured around the same binaries of pure, good, and innocent (indigenous) nature and evil (settler) invaders. However, unlike the vilification of 'invasive' non-indigenous animals in the bilby stories, the villainous invaders in the spirit bear stories are all human. Moreover, they are not just individual bad humans invading and destroying the environment, but powerful operatives from industry and government. By being invited to go into a battle with corporate greed and government complicity (not just with rabbits) the young audiences of these Canadian texts are being interpolated into an explicitly political activist agenda.

The text that exemplifies all of this is the well-known film *Spirit Bear: The Simon Jackson Story* (2005), written by Kent Staines, directed by Stefan Scaini, and adapted as a television series, which offers a fictionalised account of teenage Simon Jackson's impassioned crusade to build a youth movement to save the spirit bear from likely extinction. It shows how Jackson bravely confronts and challenges the corporate greed of the multimillion-dollar forestry industry that is logging the Great Bear Rainforests and destroying the spirit bear's habitat and begins to lobby government. One of the key themes is the successful way that Jackson actively inspires and mobilises the support of his peers, who increasingly band with him in this 'David and Goliath' battle.

Although inspired by Simon Jackson's real-life story, the film takes fictional licence by adding some compelling dramatic embellishments. These include portraying young Jackson as having a very special and intimate relationship with a spirit bear who saves him from an attack by another animal in the rainforests and also with an indigenous elder, who is his wise guide and mentor. The bear engenders his love and indebtedness, and the indigenous guide teaches him traditional stories that confirm the special spiritual significance of the spirit bear. At school, he learns that the spirit bear is a rare and endangered animal. With all of this motivating him, Jackson summons the 'power of one' and transforms from an awkward shy boy to a courageous and inspirational youth leader with a big vision, which he clearly articulates in the rousing speech that he delivers to his fellow school students:

> We will not be silenced by big business. Decisions are being made right now about our future, without our input, without our consent. . . . We watch as people in power take over our natural resources. . . . The voice of our generation must be heard. It will be the voice of the next generation that will save the planet for future generations. We will not be silenced. . . . Together we will save our spirit bear, our planet.
>
> (*Spirit Bear: The Simon Jackson Story* 2005)

The portrayal of Simon Jackson's gallant efforts to save the spirit bear, and his rallying calls to others of his own generation to join him appeal to nationalist as well as environmental sentiments. The Jackson character is a consummate young Canadian hero. The film celebrates Jackson's love for the pristine British Columbian temperate rainforest environment and his courageous efforts to fight for it and the spirit bear by exercising his democratic rights. All of this is framed as a proud example of core Canadian values. This eco-nationalist pride is evident in the scene where Jackson promotes his campaign to the royal family during one of their visits to Vancouver. Jackson's essentially Canadian eco-nationalist credentials are also reinforced and validated by his close relationship to the high-profile indigenous Canadian character in the film. The audience is constantly reminded that this indigenous leader is a member of the Order of Canada.

The film takes pains to emphasise that Jackson has been mentored into the role of environmental youth leader by this indigenous leader and that his youth campaign has indigenous endorsement. However, although Jackson (and, by association, the film) is positioned as indigenous friendly, it seems quite oblivious to its own neocolonialist subtexts. The first of these is the subtext of cultural appropriation, whereby it appears quite 'natural' and unproblematic for a young white-settler boy to attain honorary indigenous status. This occurs when Jackson is entrusted with the spirit bear creation story and encouraged to teach it to his non-indigenous peers (who were previously unaware of the presence of indigenous peoples in the province). One of the most disturbing moments of the film occurs when Jackson realises that he will not be able to save the spirit bear if he does not turn towards reputable science. He abandons the indigenous spirit bear story and turns to modern western science for 'proof' of the spirit bears' genetic uniqueness and hence special status. At this turning point, the subtext is that (primitive) indigenous knowledges, although worthy of respect, are ultimately only based on 'myth' (as the spirit bear creation story is referred to in the film), whereas (advanced) western scientific knowledges are based on evidential truths. The closely related subtext about the inevitable march of (superior, modern western) progress is further reinforced when, after many years of devoted activism, the indigenous leader concedes that he has been unable to protect his lands from the forestry industry and hands the baton over to young, white Jackson. This aspect of the narrative not only ensures that Jackson can take centre stage as the young white-settler saviour, but it also conveys the message that indigenous cultures and knowledges will inevitably be superseded by superior western ones. The savvy young white boy is even more heroic, as he is set to succeed where the indigenous elder could not.

The Simon Jackson story is not only a well-known film and TV series based on true-life events. Behind the scenes tells another closely related story of an impressive publicity campaign, driven by Jackson's genius at community mobilisation, as well as his genius for the kind of self-promotion that has succeeded in gaining him legendary personal status as a world change agent. This can be seen in the ways that Jackson has played a key role in the cultural construction of his own legendary, almost cult, status. For example, Jackson was the executive producer of the film about his early life. The Spirit Bear Youth Coalition, led by Jackson, promoted the film to its vast international networks and vice versa. Repeated references are made in campaign materials to *Time* magazine's prestigious naming of Simon Jackson as one of its 60 'Heroes for the Planet', also mentioned on Jackson's personal website 'Ghost Bear' (Jackson 2018).

By 2016, when the survival of the spirit bear was finally ensured by the passing of the Great Bear Rainforest Land Use Order (British Columbian Gazette 2016) and Jackson blogged his declaration that the spirit bear was 'not just safe, but saved' (Jackson 2016), he had already moved on to a bigger world stage. He had co-launched a new network called CoalitionWILD (2013) to build a new global youth movement of 'rising leaders' to create a 'wilder world'. On its website, it claims that its membership base of 10,000 is providing 'a new voice for nature' (CoalitionWILD 2013).

The fact that *Spirit Bear: The Simon Jackson Story* (2005) narrates the early years of a legendary environmental activist in the making and that this story cannot be separated off from Jackson's huge success at promoting himself as this legend, highlights one of the biggest paradoxes of environmental stewardship and salvation narratives. This is their propensity towards anthropocentric grandiosity – to become more about the heroism of the human saviours than about the animals themselves. In other words, all of the hype about Jackson as an extraordinary environmental activist may have galvanised a highly successful environmental campaign, but at the cost of reifying (settler) human exceptionalism, displacing indigenous custodians, and upstaging the extraordinariness of the forests and the spirit bears themselves.

The cultural politics of settler environmental stewardship

Despite best intentions and belying the apparent innocence of aligning uncorrupted children with 'good' indigenous nature, moves within these settler eco-texts to promote young people as environmental saviours and stewards are implicated within the messy cultural politics of settler colonial relations. In our discussions of endangered bilby and spirit bear children's texts we have noted paradoxes and lacunae that point to some of the reasons why cultivating/educating settler children to be the next generation of native species stewards in Australia and Canada is neither a self-evidently good move nor a straightforward matter. There are broader political contexts and implications that require consideration. The first is that the underlying premise that human intervention-for-the-better is always necessary has become increasingly problematic in the game-changing era of the Anthropocene. The second is that unproblematised notions of young settler environmental stewardship sit very uneasily within the fraught politics of settler colonial nations – particularly in light of the interrelated colonial legacies of ecological destruction and indigenous dispossession.

As we discussed in the introduction, the declaration of the Anthropocene as the new epoch in which certain human activities have fundamentally altered the earth's biospheric and geospheric systems has blurred the distinction between human and natural forces. This realisation has also triggered a plethora of circumspective debates within the social sciences and humanities about what it means to be human and our place and agency in the world, particularly in relation to the mass anthropogenic species extinctions over the last 200 years that are so closely associated with colonisation. We concur with the critical Anthropocene scholars who call for radical rethinking that breaks with the western knowledge traditions and practices that separate humans off from other species and hence from the rest of the world, and point to the delusional short-sightedness of continuing to privilege exceptional human agency (see, for instance, Gibson *et al.* 2015). We also ascribe to the feminist critique and rejection of heroic responses to the Anthropocene (Haraway *et al.* 2015; Stengers 2015; Tsing 2015).

Unreconstructed western notions of environmental stewardship trouble us because they reiterate the structuring nature/culture binary, rather than seeking alternatives to it. Contrary to what is needed in response to the Anthropocene, conventional notions of stewardship do not contribute to radically rethinking our place and agency in the world. This is because environmental activist discourses, particular those informed by wilderness protection discourses, tend to valorise nature and seek to protect some exteriorised and pure notion of it from the corrupting and contaminating human (cultural/technological) influences. In so doing, they reverse, but nevertheless still rehearse, the nature/culture divide. Furthermore, by mobilising and promoting the rather naïve notion of (good human) stewardship as the only way of ensuring that (pure) nature is protected from (corrupting) humans and their invasive (impure) species, environmental discourses privilege human agency as pivotal to preserving and protecting valorised and revalued nature. Consciously or not, appeals for environmental stewardship not only acknowledge that (bad) humans have altered, damaged, or destroyed natural environments, but they also assume that only (good) humans have the requisite exceptional capacities to ensure the survival of exteriorised natural environments. In other words, they default to anthropocentric, bifurcating imaginaries that cannot conceive of a world without some form of human gallantry – without the ubiquitous (western, masculinist) heroic script of humans-to-the-rescue.

This heroic figure of the settler environmental steward bears very little resemblance to indigenous modes of custodianship. The very assumption that settlers and indigenous people might automatically share understandings about caring for the land and its creatures either denies or ignores the fact that not all cultural knowledge systems are structured around the nature/culture divide. Granted, it is very challenging for those of us thoroughly schooled in western dualisms to think about the human species as already a part of the (natural) world, not separate from it. However, for those who have been acculturated within indigenous cosmologies, human lives are an integral part of a relational web of beings, entities, and forces, not separate from the rest. It is a mistake to elide western stewardship notions, which typically presume some kind of detachment (as evidenced by terms such as 'resource management' and 'wildlife rescue'), with those of indigenous people, whose custodianship is about fulfilling mutual obligations and responsibilities to the land or 'country' and its creatures because all life is interrelated and interdependent. There is no bifurcation, no human heroicism in this version of custodianship.

Within the new breed of eco-nationalist texts for young people in settler colonial contexts, the inclination to position settler children as superseding indigenous caretakers is concerning. It is worrying when the explanation provided for such a positioning infers that it is the *absence* of the 'original people' (as in the opening quotation) that necessitates children taking up the stewardship baton. It is equally worrying when indigenous custodianship is recognised and affirmed but then appropriated by young white environmental activists, as in the film *Spirit Bear: The Simon Jackson Story* (2005).

Moves to disappear indigenous peoples from the land, either by relegating them to the distant past (for example in Bright's *Burra Nimu*) or by defaulting to sanctified wilderness tropes and appropriating indigenous people's knowledges (as in Staines' and Scaini's *Spirit Bear: The Simon Jackson Story*) have been well critiqued by indigenous scholars. Marcia Langton (2012), for instance, speaks of 'the conceit of wilderness ideology' within the settler Australian nature conservation movement, which verges on a neocolonial form of 'terra nullius' – the fallacious 'empty land' legal doctrine that the colonists first used to claim Australian lands as part of the British Empire. And, as Laura Hall (2015, p. 288) reminds us, 'the divide between humans and the "natural" world cannot be understood in the Americas without contextualizing its origins in the Eurocentric project of genocide, ecocide, and control over Indigenous Peoples and Indigenous lands'.

Not only are the erasure of indigenous custodians and the appropriation of indigenous knowledges part of the dispossessing strategies of settler colonialism, and fundamental to claims to settler sovereignty, but ongoing settler appropriations of indigenous lands and identification with them is part of the never-completed process of naturalising settler claims to belong (Veracini 2010). Not surprisingly, settler children have been central to the process of establishing belonging through naturalisation. Securing settler children's affections for appealing native animal fictional characters has been a key strategy of this naturalisation process, as it cultivates identification with the colonised lands and their creatures from an early age (Taylor 2014).

What is alarming about the new trend, particularly in the eco-nationalist brand of *endangered* native animal children's narratives, is that it brings these well-established settler cultural practices into sharper alignment with the heroic and bifurcating tendencies of settler environmental politics. Eco-nationalist narratives risk articulating a new form of neocolonialism. By positioning settler children as the 'natural' protectors and saviours of endangered indigenous animals and environments, they inadvertently continue the process of indigenous displacement and dispossession. They also miss out on opportunities to learn from inseparable indigenous ontologies, which have much to offer in the face of the massive species losses directly caused by the man-over-nature modernising project of colonial settlement.

So what are the alternatives to such neocolonial stewardship positionings and disavowals? Is it possible to deliver an ecological message without resorting to a simplistic formula of goodies and baddies? How might we reconceptualise settler ecological care in ways that do not inadvertently rehearse colonialist conceits and draw upon tired old scripts of heroic settlers to the rescue? What might a decolonising ecological ethics look like?

Towards a decolonising ethics of ecological reconciliation

To circumvent the predetermined moral binary codes and the tropes of (settler) human exceptionalism and heroicism that we have observed in the Australian children's eco-literature and in the Canadian film, we respond to Deborah Bird Rose's (2004, 2011) consistent calls for reconnecting and reconciling instead of rehearsing separations within a broader decolonising ecological ethics.

Combining philosophical and ethnographic modes of inquiry and focusing upon the ethical questions raised by multispecies cohabitation in this age of anthropogenic mass extinction (Rose 2011; van Dooren 2011, 2014), Rose has consistently called for a 'decolonising ethic' (Rose 2004) to envision new kinds of ecological relationships. Her thinking is significantly informed by her Aboriginal Australian 'teachers', the people she worked with as an anthropologist in the Northern Territory. Taking its lead from inseparable indigenous onto-epistemologies, Rose's reconciliation is not just about atoning for past colonial injustices that settlers inflicted upon indigenous people and tackling the current inequities that flow from these, as is the focus of settler colonial governments' reconciliation policies. Most significantly, it additionally requires decolonising settler modes of thinking, acting, and relating to all other lifeforms.

Rose takes pains to explain that a decolonising ethic is essential to her notion of reconciliation, and it is a very 'strong' meaning that she is eliciting 'against domination it asserts relationality, against control it asserts mutuality, against hyperseparation it asserts connectivity, and against claims that rely on an imagined future it asserts engaged responses to the present' (Rose 2004, p. 213). Positioning her calls for 'eco-reconciliation' (Rose 2011, p. 59) within the context of the current era of anthropogenic loss and extinction, she emphasises the urgent need to fully appreciate something that indigenous people already know – that we share a common kinship with all other species and that all life and death is interconnected. This understanding is congruent with the ecological worldview that stresses that 'all living things are bound up in the web of exchanges that makes life possible' (Rose 2011, p. 60).

This recognition that human lives are inextricable entangled with those of other species underscores our precarity in the Anthropocene. As Rose points out, 'Our interdependence is our blessing and our strength. But as we are increasingly beginning to understand because it is all falling apart, our interdependence is also our peril' (Rose 2011, p. 61). With their inherently human-centric premises, western ethical frameworks are inadequate to the task of redressing these perils. They produce halfway responses, such as affirmations of the need for human environmental stewardship, that are limited because they remain ultimately entrapped by their own binary solipsisms. In seeking alternative pathways for recuperation at this unprecedented time of death and loss, Rose reflects upon Australian Aboriginal people's tradition of 'singing up the country' – a ritual practice that is undertaken in order to reboost, revitalise, and reaffirm the life-giving exchange of forces that come from the land and all its creatures. Singing up country is a replenishing enactment of encounter and relationality to ensure mutual prosperity. Drawing upon this notion for renewing connectivities and thereby kickstarting recuperation, Rose asserts that 'eco-reconciliation' must be about 'living generously with others, singing up relationships so that we all flourish' (Rose 2011, p. 59).

A decolonising ethic of ecological reconciliation interrupts the conventional wisdom of settler stewardship narratives that seek to position children as the protectors or saviours of endangered native animals and 'pristine' natural wilderness areas. It interrupts the premises that nature is 'out there' passively waiting for us to look after it. It interrupts the settler 'conceits of wilderness ideology' (Langton 2012) and of human exceptionalism and heroicism. It interrupts the colonialist logics of detachment and separation, of opposition and conflict. Choosing ecological reconciliation over settler stewardship entails raising the next generation to recognise that our ethical responsibilities are based, not upon our 'specialness' as a species, but upon the simple fact that we are always already inextricably bound up with other species within the common lifeworlds that we all inherit. The job of a decolonising ethics of ecological reconciliation, therefore, is to redouble our efforts to renew and revitalise our worldly relations and responsibilities, not because we are a superior and potentially benevolent species, but because we are implicated as one amongst many, and our fates and futures are inexorably bound together.

References

Barcott, B., 2011. Spirit bear. *National Geographic Magazine*. Available from: http://ngm. nationalgeographic.com/print/2011/08/kermode-bear/barcott-text [Accessed 9 February 2018].

Bright, J., 2012/1993. *Burra Nimu, the Easter bilby: An Australian children's story*. J. Selby, Illustrator. Available from: http://easterbilby.weebly.com/uploads/3/1/4/4/3144138/eas ter_bilby_story_-_to_downoad_1_mb.pdf [Accessed 9 February 2018].

British Columbian Gazette, 2016. Notice: New land use orders for the Great Bear Rainforest, 156(4), 28 January 2016. Available from: www.bclaws.ca/civix/document/id/bcgaz1/bcgaz1/1938462071 [Accessed 9 February 2018].

Burra Nimu, the Easter Bilby, n.d. *Website*. Available from: http://easterbilby.weebly.com/ [Accessed 9 February 2018].

CoalitionWILD, 2013. *Website*. Available from: https://coalitionwild.org/ [Accessed 9 February 2018].

Eco Books 4 Kids, 2013. *The book: Spirit Bear*. Available from: https://ecobooks4kids. wordpress.com/spirit-bear-childrens-book/ [Accessed 9 February 2018].

Flannery, T., Kendall, P. and Wynn-Moylan, K., 1990. *Australia's vanishing animals*. Sydney: RD Press.

Rabbit Free Australia, 2017. *Website*. Available from: www.rabbitfreeaustralia.com.au/ [Accessed 9 February 2018].

Garnett, A. and Kessing, K., 2006/1994. *Easter bilby*. Alice Springs, Australia: Kaye Kessing Productions.

Gibson, K., Rose, D.B. and Fincher, R., eds., 2015. *Manifesto for living in the Anthropocene*. Brooklyn: Punctum.

Haigh's Chocolates, n.d. *The bilby*. Available from: www.haighschocolates.com.au/about-us/the-bilby/ [Accessed 9 February 2018].

Hall, L., 2015. My mother's garden: Aesthetics, indigenous renewal, and creativity. In: H. Davis and E. Turpin, eds. *Art in the Anthropocene: Encounters among aesthetics, politics, environments and epistemologies*. London: Open Humanities Press, 283–292.

Haraway, D.J., 2016. *Staying with the trouble: Making kin in the Cthulucene*. Durham: Duke University Press.

Haraway, D.J., Ishikawa, N., Gilbert, S., Olwig, K.R., Tsing, A.L. and Bubandt, N., 2015. Anthropologists are talking – About the Anthropocene. *Ethnos*, 81(3), 535–564. doi:10.1080/00141844.2015.110583.

Harrington, J. and Arnott, M., 2013. *Spirit bear*. Toronto: Eco Books 4 Kids.

Heise, U.K., 2016. *Imagining extinction: The cultural meanings of endangered species*. Chicago: University of Chicago Press.

Hilton, N. and Whatley, B., 2006. *The smallest bilby and the midnight star*. Sydney: Working Title Press.

———., 2009. *The smallest bilby and Easter games*. Sydney: Working Title Press.

———., 2014. *The smallest bilby and the Easter tale*. Sydney: Working Title Press.

Howlet, M., Rayner, J. and Tollefson, C., 2009. From government to governance in forest planning? Lessons from the case of the British Columbia Great Bear Rainforest initiative. *Forest Policy and Economics*, 11, 383–391.

Jackson, D.S., 2016, 3 February. *Spirit bear saved*. Available from: http://ghostbear.co/spirit-bear-saved/ [Accessed 9 February 2018].

———., 2018. *D. Simon Jackson: The next journey*. Available from: http://ghostbear.co/d-simon-jackson/ [Accessed 9 February 2018].

Kessing, K., 2015a. *Bilby's ring: Book 1, Out of the spinifex*. Alice Springs, Australia: Kaye Kessing Productions.

———., 2015b. *Bilby's ring: Book 2, Across a great wide land*. Alice Springs, Australia: Kaye Kessing Productions.

———., 2015c. *Bilby's ring: Book 3, Into the bowels of the biggest city*. Alice Springs, Australia: Kaye Kessing Productions.

Langton, M., 2012. *The conceit of wilderness ideology*. Boyer Lectures (Lecture 4), Australian Broadcasting Commission. Available from: www.abc.net.au/radionational/programs/boyerlectures/2012-boyer-lectures-234/4409022 [Accessed 9 February 2018].

McAllister, I. and Read, N., 2010. *The salmon bears: Giants of the Great Bear Rainforest*. Victoria: Orca Books.

Rabbit Free Australia, 2017. *Website*. Available from: www.rabbitfreeaustralia.com.au/ [Accessed 9 February 2018].

Ritland, K., Newton, C. and Marshall, D., 2001. Inheritance and population structure of the white-phased 'Kermode' black bear. *Current Biology*, 11, 1468–1472.

Rose, D.B., 2004. *Reports from a wild country: Ethics for decolonisation*. Sydney: UNSW Press.

———., 2011. *Wild dog dreaming: Love and extinction*. Charlottesville: University of Virginia Press.

Save the Bilby Fund, 2017. *Website*. Available from: www.savethebilbyfund.com/ [Accessed 9 February 2018].

Sibley, I., 1994. *The bilbies' first Easter*. Melbourne: Silver Gum Press.

Spirit bear: The Simon Jackson story, 2005. Film. Written by K. Staines, directed by S. Scaini. Toronto: Screen Door.

Spirit Bear Youth Coalition, n.d. *About*. Facebook webpage. Available from: www.facebook.com/pg/SpiritBearYouth/about/?ref=page_internal [Accessed 9 February 2018].

Stengers, I., 2015. Accepting the reality of Gaia: A fundamental shift? *In*: C. Hamilton, C. Bonneuil and F. Gemmene, eds. *The Anthropocene and the global environmental crisis: Rethinking modernity in a new epoch*. London: Routledge, 134–144.

Taylor, A., 2014. Settler children, kangaroos, and the cultural politics of Australian national belonging. *Global Studies of Childhood*, 4(3), 169–182.

Threatened Species Network, n.d. *Australian threatened species: Greater bilby Macrotis Lagotis*. Fact Sheet. Available from: www.environment.gov.au/system/files/resources/54adf0bf-37a7-44e7-ada6-f60e9c7481d4/files/tsd05greater-bilby.pdf [Accessed 9 February 2018].

Tsing, A.L., 2015. *The mushroom at the end of the world: On the possibility of life in capitalist ruins*. Princeton: Princeton University Press.

van Dooren, T., 2011. Invasive species in penguin worlds: An ethical taxonomy for killing in conservation. *Conservation and Society*, 10(4), 286–289.

———., 2014. *Flight ways: Life and loss at the edge of extinction*. New York: Columbia University Press.

Veracini, L., 2010. *Settler colonialism: A theoretical overview*. London: Palgrave Macmillan.

Wright, K., 2014. An ethics of entanglement for the Anthropocene. *Journal of Media, Arts, Culture*, 11(1). Available from: http://scan.net.au/scn/journal/vol11number1/Kate-Wright.html [Accessed 9 February 2018].

———., 2017. *Transdisciplinary journeys in the Anthropocene: More-than-human encounters*. New York: Routledge.

Zielinski, S., 2012. *Chocolate bilbies, not bunnies, for an Australian Easter*. Available from: www.npr.org/sections/thesalt/2012/04/04/150005331/chocolate-bilbies-not-bunnies-for-an-australian-easter [Accessed 9 February 2018].

5 Children, raccoons, and possums
An ethics of staying with the trouble

> Contact zones . . . are where assemblages of biological species form outside their comfort zones.
>
> (Haraway 2008, p. 218)
>
> Contact zones are full of the complexities of different kinds of unequal power that do not always go in expected directions.
>
> (Haraway 2008, p. 218)

Child-wildlife cohabitation in the multispecies contact zone

The children-raccoon-possum common world assemblages that are the subject of this chapter take shape within what Donna Haraway (2008, pp. 217–218) refers to as the discomforting multispecies 'contact zone'.[1] We see the contact zone as an inherently unsettled and unsettling interstitial semiotic-material space character-ised by the effects of (colonial) displacement and by the consequential 'thrown-togetherness' (Massey 2005) of unlikely multispecies partners. The assemblages that form within multispecies contact zones, such as those between urban-dwell-ing raccoons and children and urban-dwelling possums and children, have noth-ing to do with choice or compatibility. They are not assemblages of mutually sustaining companion species relations or symbiotic lifeforms. Instead, these characteristically disconcerting 'contact zone' assemblages constellate somewhat randomly through the intersecting trajectories of species 'on the move', displaced by the 'power geometries' (Massey 2005) of settler colonialism, global migra-tions and diasporas, rapid urbanisation, deforestation, and habitat destruction. As we discuss in some detail in the introduction, the multispecies common worlds in which we all live are the product of such constellating power relations. Each has its own distinctive and shifting sets of natureculture assemblages, thrown together more by our entanglement in messy (colonial and ecological) inheritances than by intention, natural order, or choice.

Unlike the children's real-life convivial encounters with shy kangaroos and deer in Chapter 2 and with the small but powerful ants and worms in Chapter 3, in this chapter, it is the boundary-crossing behaviours of raccoons and possums, and the subsequent uneasiness about children cohabiting with them at close quarters, that distinguish these as discomforting 'contact zone' relations. In the sections that follow, we describe how the unruly raccoons and pesky possums, respectively, that move into urban areas and set up residence in early childhood centres in Canada and Australia breach proximate comfort zones and unsettle power relations. The determined boundary-crossing behaviours of these brazen wild native animals subvert the divisive colonial orderings that seek to partition off urban-dwelling humans (and their selected 'pets') from native wildlife, which is expected to reside 'out there' in the natural environment. They cause the (settler) humans to struggle to maintain their desired but ultimately impossible boundaries.

In the accounts that follow, we detail the cultural, species, and geographical specificities of both sets of urban-child-native-wildlife-residential assemblages. We trace the spectrum of mixed, awkward, and discomforting effects that these similar but not exactly the same assemblages produce – at least for the adults and children involved. We highlight the ethical tensions and complexities that are thrown up when children cohabit with native wildlife in the urban contact zone and engage with Haraway's (2016) encouragement for us to 'stay with the trouble' in the multispecies 'contact zone' as a form of ethical response.

The raccoon-settler contact zone

Raccoon inheritances

The raccoon has always been North America's most enigmatic animal. This is because of its wily intelligence and its complete lack of fear of humans – a combination of attributes that has variously puzzled, bemused, confounded, impressed, guided, and threatened the humans with whom it cohabits.

Amongst North America indigenous peoples, the raccoon is held in highest regard. Its intelligence is often depicted as outwitting that of humans. Many First Nations and Native American creation stories characterise the raccoon as a trickster figure who not only outsmarts humans but is a kind ironic muse – enacting teachings about respect, cooperation, honesty, and hard work. In indigenous cosmologies, raccoons, as all other native animals, are closely related to humans and convey all sorts of important lessons about living well together (Elliott n.d.; H.P. Taylor 1996). For instance, spiritual teacher Bobby Lake-Thom (1997), who is of Karuk and Seneca descent, writes that raccoons are used as 'doctor power, hunting power and protection power', helping people to find their way when lost or to resolve a problem.

Settler-raccoon relations have been somewhat more fraught. During the early colonial days, raccoons were heavily hunted for their pelts, which were traded throughout the 18th and 19th centuries. Despite these concerted assaults, raccoons not only survived but have flourished. However, their patterns of life have dramatically changed. Urban raccoon populations have grown exponentially in the last few of decades, so much so that most raccoons now live in close proximity to humans (Nowak 1991; Pettit 2010; Raccoon Nation 2014). The concentration of raccoon populations in urban environments is closely associated with industrial agriculture and forestry, habitat loss and the alternative food sources provided by human settlement, and also with this species' remarkable ability to adapt. As efficient mesopredators, raccoons are reputed to be one of the best urban-adaptable species on the planet. Their keen intelligence and apparent fearlessness makes them particularly adept at exploiting human resources, especially in large North American cities such as Vancouver, Toronto (the raccoon capital of the world), New York, and Chicago (McKinney 2002; Bozek *et al.* 2007). In a recent documentary about raccoons, pointedly named *Raccoon Nation* (2014), David Suzuki refers to raccoons as 'urban sophisticates' and points out that it is actually humans who are 'pushing [raccoons'] brain development and perhaps even sending them down a new evolutionary path'. Unlike Suzuki's clear respect for the raccoons' level of sophistication and adaptive intelligence, wily raccoons are typically feared and despised in urban North America. In fact, it is largely because of their brash intelligence and their clever modes of adapting to urban life that raccoons are widely regarded as invasive pests posing a direct threat to human safety (Corman 2011; Nagy and Johnson 2013).

Raccoons' brash intelligence and adaptivity also present significant epistemological challenges to humanism's binary logics. Where do they fit in to the biological schema in which highly reasoned intelligence and the capacity to learn are seen as exclusively human and therefore apex attributes? Psychologist Michael Pettit (2010) traces the dilemmas that raccoons have posed to North American scientists, starting back in the 18th century, when natural scientists spent decades debating where to place raccoons in the natural order and on the hierarchy of beings in relation to 'Man'. He notes that the raccoon's intelligence, curiosity, and mischievousness presented much taxonomic confusion as to whether they were wild animals, vermin, or companions. This confusion carried on into the early 20th century, when raccoons were declared to be ideal household pets because of their 'charm' and ability to live with and adapt to humans, and were also introduced into science laboratories so their intelligence could be studied. Neither of these experiments lasted long. Scientists soon retracted their previous advice and warned the public that raccoons were too difficult to serve as pets when the animals' outrageously unruly behaviour (opening doors, stealing food, escaping, and eluding capture) caused domestic chaos. Similarly, the raccoons' unruly behaviour in the laboratories (tearing their cages apart, escaping, and biting scientists) quickly led researchers to realise that the raccoons were noncompliant experimental subjects and too risky to hold in confinement. They did glean, however, that raccoons have a form of reasoning ability that is very similar to humans' (Pettit 2010).

Raccoons in popular culture

In children's popular culture, representations of fictional raccoon characters as smart, charming, and curious but also fearless, aggressive, and conniving continue to reflect the spectrum of ambivalent feelings they evoke and the ways in which they seem so uncannily similar to humans. Given how disconcerting they are, it is not surprising that raccoons are always the active and never the passive characters. Rather, through their portrayal in cultural texts, raccoons have attained iconic status as archetypal troublemakers. Examples abound. For instance, in the children's novella *Sammy Squirrel and Rodney Raccoon to the Rescue* (Lawrence and Clover 2011), the squirrel and racoon, who live in Vancouver's Stanley Park, embark on an adventurous plan to rescue their friend Judy Crow, who had been kidnapped by 'crow-nappers'. Both Sammy Squirrel and Rodney Raccoon are depicted as intelligent and inquisitive animals as they intentionally and carefully plan Judy Crow's rescue by reading books about crows' behaviours. Yet, unlike Sammy Squirrel, Rodney Raccoon is also a mischief-maker. He continually gets into trouble and derails the rescue plans because of his insatiable hunger and his escapades to steal human foods. Similarly, but with a much more violent twist, the central raccoon character in the computer-animated feature film *The Nut Job* (2014) is a hardened troublemaker. Norvirus Raccoon is the self-appointed ruler of Liberty Park and a violent, deceitful, calculating thief who is prepared to kill for food. Raccoon leads a mob of park animals whose task is to steal and store food for the winter inside a massive tree in a park. His carefully planned and military-style food raids end with a deadly fight against his despised enemy, Surly Squirrel.

These typical narrative depictions of raccoons in children's popular culture not only reinforce the long and widely held perception of raccoons as threatening because they are both highly intelligent and mischievous animals, but they also feed a widespread sense of anxiety about encountering actual raccoons in everyday life. In British Columbia's leafy urban contexts (the setting of both of these raccoon narratives and of Veronica's ethnographic research), real-life encounters between cohabiting children and raccoons are inevitable. Moreover, these commonplace real-life encounters cannot be disentangled from the uneasy raccoon discourses and cultural representations that surround them. They are always affected, in some way, by the raccoons' notoriety. Through studying some of these encounters,[2] we are interested in seeing whether it is possible for the relations between children and raccoons to exceed the discursive anxieties that surround them.

Unruly raccoons

The mother raccoon climbs down from a tree with her two cubs. Crossing through the sandbox, she enters a small tent in the playground, a favourite domestic play space for the children. The cubs follow her. Soon, the whole family is frolicking in and around the tent, entertaining a group of nearby children who stand watching with wide eyes and open mouths. The mother raccoon moves two red cushions from the tent as if rearranging the furniture, and immediately her two kits sit on them. As uncannily human-mimicking 'homemakers', they appear to be extremely comfortable in the children's playground. The mother racoon places one more cushion beside the kits for herself. She sits on it and watches as her kits playfully manipulate the plastic plates, cutlery, mugs, and other toy kitchenware the children have left in the tent.

Despite the charming domesticity of this scene, the raccoons' new residency in this British Columbian childcare centre has caused much disquiet. As unpredictable wild native animals, they are regarded as unwelcome intruders in this carefully managed urban environment, crossing the spatial and ontological boundaries between wild and domestic, between human and animal. Even more provocative is the raccoons' unruly behaviour. These mischievous 'masked bandits' show no respect for human property. They sabotage carefully planted garden boxes, vandalise trashcans, rearrange the playground, tap on the classroom skylights, and play with the children's toys. They defy human authority, sabotage human plans, and deviate from expected paths. In short, they live up to their reputation as a dangerous threat because they unsettle the (colonial) binary order, which is based upon a clear separation between 'civilised humans' and 'wild creatures'. As Lesley Instone (2014, p. 83) describes in discussing other unruly wild urban species, the raccoons are seen as 'troublemakers' who 'confound human expectations and dominant norms of urban landscape . . . and bodily orientations'. This disconcerting urban-child/wild-raccoon residential assemblage turns the childcare centre into a multispecies contact zone, prompting the responsible adults to introduce new routines, reorient behaviours, and adjust their own and the children's patterns of movement in the playground (see Figure 5.1).

An educator calls the children to join her in the playground. She has carefully executed the routine raccoon check and determined that it is safe for the children to come outside. She warns the children not to go near the shed, however, because the raccoon family could still be hiding there. The children nod and move

Figure 5.1 The children watch the raccoon in the playground.
Author's (Veronica Pacini-Ketchabaw's) photograph

cautiously outside to play. All of a sudden, a child sounds the prerehearsed alarm: 'Raccoon, raccoon, raccoon!' She has spotted three raccoons climbing down the tree closest to the building. Before the animals can touch the ground, the educator clusters the children together on the opposite side of the playground, maintaining the prescribed 'safe' distance from the raccoons. Panicked, she leads the children to the classroom door, but the raccoons reach the door first. It seems too risky now to enter the classroom, so the educator leads the frightened children back to the spot where they first gathered and announces that they will not be able to play or go inside until the raccoons leave the playground.

Inside the classroom, another educator is preparing the tables for the children to have lunch. She hears the commotion outside, and when she investigates, she sees the raccoons are approaching the classroom door. She runs to close it and also ensures that the windows are sealed. Because the raccoons can easily move from playground to playground along the fence, this educator calls to warn her colleagues in the adjoining classroom. They quickly gather the garden hose in case they need to use water to fend off the intruders. The raccoons, though, are still in front of the original classroom door, and the children remain huddled on the opposite side of the playground.

The educators' struggles to maintain a sense of order and separation in this multispecies contact zone are not just spatially enacted. Raccoons not only cross physical boundaries when they enter the playground, but they also threaten to breach the biologic human/animal divide. As 'trash animals' (Zahara and Hird 2015), they are susceptible to carrying infectious diseases, such as rabies, leptospirosis, or salmonella, that can potentially be transferred to humans. The educators are particularly worried about the 'raccoon worm' known as *Baylisascaris procyonis*, which develops in the raccoon intestine, producing millions of eggs that are passed in the animal's faeces and can survive in the environment for several years (Gavin *et al.* 2005). Humans are infected if they ingest fertile eggs, and because children frequently put objects, hands, and dirt into their mouths, they are deemed to be at particularly high risk of ingesting infectious raccoon poo. Despite the fact that raccoon droppings have rarely been found in the playground, every morning and every afternoon, an educator puts on disposable gloves, carries a small shovel and a garbage bag, and diligently heads out into the playground to conduct a 'poo check' before the children go outside. The possibility of children coming into contact with infectious raccoon faeces unsettles the normalised western understandings of children's spaces as needing to be hygienic and thus free of bacteria and disease.

British public health sanitisation discourses travelled to the colonies in the 19th century (Edmonds 2010), and in these 'uncivilised' lands, they took on a particular relationship to 'wild nature'. In the colonies, sanitation became entangled with the perceived need to maintain a clear divide between the (racial and cultural) 'purity' of the white settlers and the contaminating influence of the 'filthy', 'savage' native wildlife. These colonial sanitisation discourses can be traced into the present-day mandated hygiene requirements of North American early childcare providers and at least partially account for the reasons why the presence of unruly, wild, and potentially infectious raccoons in the childcare centre causes so much panic.

With her two front paws, the mother raccoon picks up the plastic bucket filled with water that the children left in the middle of the sandbox where they were building sandcastles. She carries the bucket to her kits, who are waiting for her at the sandbox ledge. As soon as she puts the bucket on the ground, the kits dip their paws in the water and start splashing in the same way the children often do when they play with water.[3] The mother raccoon watches as if supervising them. After a few seconds, the kits shift to what seem like much more intentional actions. Using their front paws, they appear to wash their faces.

Captivated and charmed by the raccoons' humanlike behaviours, a group of children and educators watch through the classroom window. All of a sudden, the mother raccoon turns her head toward the window as if she is letting these curious humans know she is aware of them. Responding to this motion, one of the children places his hand on the window to gesture hello. The mother raccoon leaves her kits and approaches the window. Without hesitation, she raises her paw to meet the child's hand through the glass. Silently, the child and the raccoon gaze into each other's eyes. The other children and the educators look at each other with surprised expressions. No one moves until the raccoon walks away from the window and rejoins her own young.

Witnessing such beguiling raccoon behaviours and intimate encounters creates a new kind of disconcertment. The children and educators are impressed by the raccoons' cleverness and inquisitiveness, entertained and amused by their playful antics, and moved by their invitations to engage. The children seem drawn to identify with the kits, who are behaving and playing in ways very similar to their own. The educators also affectionately note the similarities between the young of both species. However, all these warm emotions stand in stark contradiction to the prevailing negative social discourses about raccoons and the kinds of fearful responses they evoke. Stirred by close-up encounters and the unfolding of real-life relations, such an affection towards the raccoons is not commensurate with the panic that is induced by the enactment of raccoon emergency drills, nor with the repulsion that is felt around the possibility of being contaminated by infectious raccoon faeces. Perhaps the most disconcerting aspect of these raccoon-child relations is the awkward and ambivalent mix of affects that is always 'bubbling in and out' (Lorimer 2014, p. 196) in this multispecies contact zone.

The possum-settler contact zone

Possum inheritances

There are parallels between possum-settler relations in Australia and raccoon-settler relations in North America, as possums also cross the domesticated human/wild animal boundary by moving into human spaces (Power 2009; O'Sullivan *et al.* 2014). Possums are fondly represented within the archive of Australian children's popular culture, most famously in Mem Fox and Julie Vivas's award-winning

Possum Magic, first published in 1983 and now in its 30th anniversary edition. However, at the same time, real-life possum-settler relations are under increasing pressure because of the intensities and challenges of urban cohabitation. In other words, possum-settler relations, like raccoon-settler relations, unfold within the complex and shifting power relations of the urban human-wildlife contact zone. However, as well as the similarities, there are some marked differences between these two contact zones, attributable to the nature of the animals themselves and to the specificities of North American and Australian indigenous and settler cultural histories.

Like the vast majority of Australian mammals, possums are marsupials. They are also nocturnal, arboreal, and herbivorous. They sleep during the day in elevated, sheltered spaces and venture out at dusk to eat leaves, fruit, and flowers. There are many different kinds of possums, but to use the western scientific vernacular, it is only the two most 'common species', the common brushtail possum (*Trichosurus Vulpecula*) and the common ringtail possum (*Pseudocheirus Peregrinus*), that cohabit with humans. Of these, the ringtails are the smaller and more social. They live along the eastern and southern coastal regions in family groups, constructing raised, spherical nest-like dreys out of grass and bark. The larger brushtail possums are the most numerous and widely distributed, living in inland as well as coastal areas. Brushtails are solitary and highly territorial animals, living alone in tree hollows or similar kinds of dens (Kerle 2001).

For Indigenous Australian peoples, possums (like all native animals) are part of a complex network of interspecies kin relations, specific to the land or 'country' in which they live. For those who are born into country with 'possum dreaming', possums hold totemic and thus ontological significance. In other words, people with possum dreaming are directly related to possums. A well-known 'possum' person is Clifford Possum Tjapaltjarri, an Anmatyerre man and an internationally renowned Central Desert Aboriginal artist. His signature and famous painting 'Possum Dreaming' tells the story of his ancestral possum, 'Upamburra', travelling across the desert landscape while swishing his tail. This painting is a story of kinship and travel, tying Tjapaltjarri to the traditions of his possum ancestry and passing on his knowledge. In an interview about his artwork before he died, Tjapaltjarri spoke of his paintings as a contemporary modality for 'carrying on the Dreaming' and for 'schooling' young people (Johnson 2007).

Aboriginal peoples have highly significant and interconnected cultural, ancestral, and pragmatic relations with all animals they live with, including possums. Possum skins were of key significance to Aboriginal peoples living in the cooler southern and southeastern regions of Australia. Before and immediately after colonisation, people wore possum skins, wrapped their babies in them, used them as blankets, and were even buried in them. Such practices were gradually replaced by the adoption of western clothing and the use of woollen blankets. As Vicki Couzens (2011)[4] notes in her essay 'Kooramook Yakeen – Possum Dreaming', carefully crafted possum-skin cloaks were central to Aboriginal people's lives. As well as providing warmth and comfort, they were used in ceremonies and rituals and provided a direct connection between people, other living creatures, and country. Each panel on the hide side of a cloak, worn to the outside, was engraved with symbols depicting specific kinship and place relations. Together, the panels on the cloaks told a cumulative story of country, much like a dreaming painting, and they were similarly used as a mode of communicating people-place-animal belonging, journeying, and connectivity.

Since colonisation and the progressive destruction of their natural habitats, possum numbers have declined in rural and remote areas. A couple of the smaller possum species are currently listed as endangered. Other species are on the rise, but only in the cities. Both ringtail and brushtail possums have been highly successful at adapting to urban life. Brushtails, in particular, are ubiquitous residents of the major Australian towns and cities. In some of the established parkland inner-city suburbs of large cities, for instance in Melbourne, they are described as being in 'plague proportions' (*Possum Wars* 2013). This is because they thrive on the resources afforded by human settlement. They cannot survive without a safe, dry den, but if there are not enough tall tree hollows to accommodate them in the more built-up areas, they move into the chimneys and roof cavities of sheds and houses. If there is not enough staple food in the local native trees and shrubs, they eat exotic plants in domestic gardens and raid compost and rubbish bins. Whenever possible, they stay off the ground to avoid the threat of predatory domestic pets and being run over by cars. They emerge from their dens and dreys at night and traverse their territories across elevated pathways in search of food, scampering across tree and roof tops and adeptly navigating powerlines and fences. In short, urban possums do exceptionally well in the both the outer- and inner-city suburbs. Needless to say, human city dwellers are not always so delighted when trespassing possums pay no heed to private property and move, uninvited, to live on their roofs by day and eat their gardens by night.

Luckily for possums, they are 'charismatic' wild animals (Lorimer 2007). They are neither typically aggressive, nor particularly shy of humans. Because they are naturally curious, they characteristically stop and stare when they encounter humans at night. Their prominent, large, round eyes and their forward-facing ears give their faces an open and intelligent appearance. The brushtails have long, bushy tails and the ringtails have a thin, tightly curled prehensile tail that acts as a third 'hand'. Both species of possum carry their joeys (once out of the pouch) on their backs, 'like furry backpacks' (Sullivan 2006), and they eat their food in quite humanlike ways, by holding it in their front paws. All of these endearing characteristics offset their troublesome attributes, and they are most commonly regarded as 'cute' and harmless but nevertheless pesky wild urban animals.

White settlers never systematically slaughtered possums as they did kangaroos. This is because, unlike kangaroos, possums were not seen as a threat to farmers,[5] nor was their meat incorporated into the colonial diet. However, from the early colonial days, their soft fur was highly prized for rugs and clothing. While it is doubtful that many early British settlers were aware of the ceremonial and cultural significance of possum skins to Aboriginal people, their visible use of the skins may have influenced the colonists to follow suit. By the early 19th century, possums were commercially hunted for their skins, and this quickly turned into a profitable export industry, with increasing demand for possum-skin products in both North America and Europe. The hunting and selling of possum skins provided a supplementary form of income for those rural dwellers eking out an existence in the Australian 'bush' (Cubit 1998). Throughout the first half of the 20th century, possum-fur fashion items were in great demand, but the industry virtually collapsed after this, when the introduction of new synthetic fabrics caused a change in western fashion trends and calls for the protection of native animals from the growing Australian conservation movement started to gain political traction. By the late 1970s and early 1980s, Australian states and territories passed a series of native animal protection acts that prohibited the killing of possums, except under license and in special circumstances.

These laws have persisted, and today, a number of state environmental departments have possum management policies that aim to simultaneously protect possums and mitigate against them causing damage to property. These departments have 'catch and release' guidelines that stipulate that possums must be relocated within 50 metres of the house in which they have set up residence, as they are highly territorial and unlikely to survive outside of their home territory. If there is no alternative tree-hollow home for them, a 'possum box' must be provided and fixed in an accessible, elevated place to serve as a safe den from predator cats, dogs, and foxes. The process can only be undertaken with a license or by an already licensed commercial operator, sometimes referred to as 'possum removalists', in order to ensure that, once removed from house roofs, possums are returned, unharmed, back into their local neighbourhood (New South Wales Government Office of Environment and Heritage 2016).

Possums in popular culture

Possums occupy an unambiguously affectionate position in Australian popular culture. The word 'possum' is used as a friendly Australian idiom, commonly used by parents as a term of endearment for their children. It has been popularised by the Australian comedian Barry Humphries, whose trademark satirical welcoming audience greeting, in his stage persona as Dame Edna Everage, is a cheerful 'hello possums'.[6]

Native animal children's stories have played a key role within the cultural politics of Australian national belonging (Taylor 2014). They have been key to securing settler Australian children's identification with the 'strange' (in other words decidedly non-European) kinds of antipodean species and hence with the new nation. Because of the appeal of their soft, furry bodies and endearing big-eyed faces, possums are one of the easiest native Australian animals to anthropomorphise and to love.

There are many affectionate possum children's picture books, including *Possum in the House* (Jensen 1989) and *Possum Goes to School* (Carter 1992). Both of these books address the disruptive boundary-crossing behaviours of possums who venture into human dwellings. Both acknowledge the havoc that intrusive possums cause, but also encourage their child readers to be on the possums' side.

In *Possum in the House*, there is no doubt that the possum is much more vulnerable than threatening. He is clearly frightened by being trapped inside, and the chaos that ensues is the result of this fear. In *Possum Goes to School*, the adults are the ones who repetitively cry 'oh no!' as the possum rampages throughout the school, while the children cheer the possum on, crying out 'yes, yes!' Both stories end with the possums being allowed to stay – in a bed and in a tree in the playground.

By far the most well-loved and iconic possum story is the Australian children's picture book classic *Possum Magic* (Fox and Vivas 1983). At the beginning of the book, its anthropomorphised lead characters, Grandma Poss and young Hush, are introduced as living 'deep in the Australian bush'. Grandma Poss has magic powers, and she makes Hush invisible, so she can stay 'safe from snakes' and have lots of adventures without being seen. However, Hush is not happy being invisible. She wants to know what she looks like, and she asks Grandma Poss to reverse the spell. Unfortunately, Grandma Poss cannot do this herself, but she thinks that eating certain kinds of food might help. So they set off on a journey around Australian cities, eating iconic settler Australian foods, such as Anzac biscuits, pumpkin scones, a vegemite sandwich, and a piece of pavlova. As she eats, Hush gradually becomes visible again.

The most interesting thing about the *Possum Magic* narrative is that young Hush becomes visible when she moves into Australian cities, abandons her native plant diet, and adopts the settler cultural fare. This is simultaneously a reversal of the classic Australian children's literature naturalising trope, whereby native animals teach settler children about the wonders of the Australian bush[7] and a story that mirrors the history of the brushtail possums' adaptation to urban life. Significantly, there is no hint of the tensions and dilemmas that arise when possums move into the cities and eat human food. There is nothing but the magical innocence of possums and a celebration of Australian children's reputedly favourite foods.

Pesky possums

'Possum wars' is term that Australian media has recently adopted to describe an unfolding series of real-life urban settler-possum contact zone events. The 'city battleground' in question is in inner-northern Melbourne, where, according to reports, 'it is raining possums' (*Possum Wars* 2013). The first to use the combative descriptor was a feature newspaper article in *The Age* titled 'The Possum Wars' (Bock 2011) and subtitled 'Furry, cute, noisy and destructive: the common possum has got Melbourne residents up in arms'. The second was a documentary film called *Possum Wars* (2013) that follows the same series of events, centred around public reactions to the 'possum plague' in Curtin Square, an inner-suburban Melbourne park, and in Clarendon Children's Centre, a nearby early childhood centre.

The newspaper article subheading confirms the mixed affects that these animals provoke. So does the documentary commentary, which notes that most people admire possums, because they are so clever and adaptive, but they are also irritated by living in close quarters with their pesky 'bad habits', particularly with their habits of urinating and defecating everywhere, lifting roof tiles, eating their way through roof fascias, and destroying gardens (*Possum Wars* 2013).

With all this talk of 'possum wars', it might initially appear that contemporary public opinions stand at odds with the pro-possum sentiments of children's popular culture. However, it turns out that both media reportings on the 'possum wars' contain large doses of irony and large degrees of respect for and goodwill towards the possums. Indeed, the 'wars' that they are referring to are not so much between the troublesome possums and the troubled humans, but between primarily 'tree-loving' and 'possum-loving' factions within the human population. No one involved is actually suggesting that the possums should be culled. They are only battling over how best to manage the possum population explosion and the pressures it is placing on the surrounding environment, including the trees. The possums are described in implicitly complimentary terms as 'urban bushrangers'[8] who care nothing about the human conflicts and tensions they are causing (Bock 2011). With tongue in cheek, the documentary narrator surmises, 'It is tough for wildlife living in cities. Human beings are hard to get along with' (*Possum Wars* 2013), and the article ends with the observation that 'possums have learnt to live with us. The only question is, are we intelligent enough to learn to live with them?' (Bock 2011).

This kind of public commentary typifies the ambivalent emotional landscape of the urban settler-possum contact zone, which grapples with the inconveniences of living with pesky possums, and with the question of how many possums are too many in confined urban spaces, but ultimately comes down in favour of the possums.

A similar scenario continues to play itself out in Canberra, at the University Preschool and Childcare Centre on the ANU campus, where Affrica conducts research.[9] Possums have always lived in the trees in the surrounding grassy bushlands, but also in the large trees in the playground area and sometimes in the roof cavities. Living at close quarters to a childcare centre is an attractive option for urban possums. When bush foods become scarce, the possums can always rely upon a steady supply of the children's discarded food scraps. They are particularly keen on the fruit. In the lean seasons, they often raid the bins during the night to get to the scraps. In Canberra's cold winters, the well-heated centre ensures a cosy roof-cavity den during the possums' daytime sleeping hours. The wooden buildings are quite old, and the project of locating and sealing all the holes in the roof is an ongoing one. There are fewer possums getting inside the building these days, but there are always unexpected events.

While walking across the roof, possums have fallen down the chimney into the cot room. One morning, staff arrived at the centre to find a baby possum had fallen through the extractor-fan lid in the bathroom and landed on the floor. Luckily s/he did not land in the toilet or s/he might have drowned. The terrified little possum was curled up next to the toilet. The staff put a 'Do Not Enter – Possum' sign on the door until a maintenance worker arrived to take the possum back outside.

Under native animal protective legislation, any possums that are removed from buildings must be reaccommodated close by. So the ANU campus maintenance crew has provided a possum nesting box high up in a gum tree above the children's sand pit (see Figure 5.2). Several possums have lived in this box over the years. The staff and children regularly look up to see if there is a possum in residence. They hope there is!

Although they are nocturnal animals, it is not unusual for possums to come out in the middle of the day, particular when the children are eating their lunch and

Figure 5.2 A possum nesting box in a gum tree high above the sand pit.

Author's (Affrica Taylor's) photograph

morning tea, and there is not a lot of bush food around. If they smell the fruit, they sometimes come bounding across the playground. But they can also creep up quietly and sit watching the children from above. It is often their dangling tails that give them away. They tend to sit in the same spot above the sand pit pergola, so the children know where to look out for them and call for their friends to gather around when they see one. It is not uncommon to see an elevated possum and a group of children underneath both staring intently at each other.

These resident playground possums are not usually regarded as a threat. To the contrary, there is always much delight when they appear, and everyone eagerly assembles to get a good look and discuss what the possum is doing. It is particularly fun when it is a mother possum with a joey on her back. As one of the educators declares, 'We love our possums'.

This is not to deny that the resident possums require the educators to do extra work; they can be inconvenient and sometimes even troublesome. For a start, possums are very messy. They wee and poo everywhere – inside the roof cavities when they get in there and in the playground on a daily basis. Sometimes when it rains, the odour of damp possum excrement wafts down from the ceiling. Once a possum died in the wall cavity, and because of the stench, the wall had to be dismantled, the possum remains removed, and the wall rebuilt. This was a particularly disruptive time.

Unlike raccoon faeces, possum poo is relatively harmless. It only contains vegetable matter and does not carry transferable diseases. However, the staff regularly sweep the playground to keep it under control. The real trouble starts when the possums' favourite blossom is in flower, and their usual pellets turn into diarrhoea. For a few weeks a year, while these flowers are in bloom, the educators have to hose down the playground every morning to clean up the mess.

The only time that the possums become really troublesome is during prolonged drought periods when food is particularly scarce. Under such stressful conditions, they are out and about much more during the day in search of food, become particularly bold, and have been known to pester and intimidate the children who are handling food. During such times, tensions increase, and it is not really possible for the children to have picnics in the playground. But these kinds of inconveniences are regarded as exceptional and/or temporary. On balance, co-inhabiting possums are seen as adding value to the children's daily experiences, and their pesky messes and occasional periods of being 'overconfident' and 'overfriendly' are accommodated with good humour.

Conclusion: an ethics of staying with the trouble

In discussing the vicissitudes of multispecies cohabitations, Jamie Lorimer (2014, p. 196) points out that it is the very 'awkwardness' of these relations that affords a new kind of 'affective and thus ethical logics'. This is because the unsettling effects of contradictory emotions work against indifference. This can be seen quite clearly in the uneasy cohabitations of unruly raccoons and children in the British Columbian childcare centre, but also in the sometimes testing but more easily accommodated cohabitations of pesky possums and children in the

Canberra context. Because these wild animal/domestic human cohabitations produce an array of different emotions, they productively 'nag' at us, as Lorimer (2014, p. 196) puts it – they are neither entirely 'comfortable, loving and caring' nor demarcated by unambiguous 'horror, abjection and phobia'. They push us towards a grappling relational ethics – an ethics that is never finally resolved, but which requires us to 'stay with the trouble' (Haraway 2016) that is continuously thrown up in the urban multispecies contact zone.

In her writings on entangled and boundary-blurring relations, Haraway (2008, 2016) considers an ethics of 'staying with the trouble' to be an important aspect of 'learning how to inherit' disordered and disorderly multispecies worlds (Haraway 2016, p. 125). She speaks about 'staying with the trouble' in the contact zone in ecologically challenging times as an ethics that requires us to take responsibility for the ways in which we have damaged local ecologies, without denying the agency or 'response-ability' of other species with whom we cohabit (Haraway 2008, p. 71). An ethics of 'staying with the trouble' requires us to persist with awkward cohabitations and learn from the ways in which we are affected by other species. It requires us to consider modes of multispecies belonging and mutual flourishing that are not contingent upon finding an ultimate solution or establishing a final peace (Haraway 2008, p. 301).

The contingencies of this never finally settled ethics relate to the 'situated' specificities of each and every multispecies contact zone. In this chapter, we have described quite different contact zone assemblages of urban-dwelling raccoons, settlers, and children in British Columbia and urban-dwelling possums, settlers, and children in Melbourne and Canberra, Australia. We have traced some of the (indigenous and settler) cultural histories that have constellated these specific assemblages. We have also considered the agency of raccoons and possums, two urban-dwelling native species that share a capacity for adaptivity and boundary crossing and a keen intelligence, but are very different animals and are not simply reducible to these attributes alone.

In short, we have been trying to articulate some similar settler child/urban wildlife assemblages and trajectories in North America and Australia without flattening out specificities and differences. The overarching point that we wish to make is that the inherently unsettled and unsettling interstitial semiotic-material space of the urban multispecies contact zone is per se a troubling space. It demands ongoing persistence and is always contingent upon the specificities and colliding trajectories of all the human and nonhuman actors involved.

Notes

1 Haraway borrowed this term from Mary Louise Pratt (1992), who originally used it to describe the radically uneven zone of colonial cultural encounter and transformation. Haraway retained Pratt's original emphasis on the unequal power relations that frame the contact zone, but extended it to include the more-than-human.

2 These stories were gathered in an early childhood centre in Western Canada where Veronica undertook a three-year multispecies ethnography (see Pacini-Ketchabaw and Nxumalo 2015).

3 The Latin name for raccoon is *Procyon lotor*, which means 'washes with hands'.
4 Couzens is a Keeray Woorroong Aboriginal artist from the southwest of Victoria.
5 See Affrica Taylor (2014) for more about the 'war' waged on kangaroos by Australian sheep farmers.
6 Although possums are held in great affection in the Australian popular imaginary, it should be noted that it is a very different situation across the Tasman Sea. In New Zealand, possums are reviled as menacing invaders (similar to the ways in which rabbits are regarded in Australia). Following their introduction from Australia, possums flourished in New Zealand and destabilised local ecological systems, leading to the extinction of a number of native species. Despite their charismatic appearance, they are effectively criminalised and seen as unequivocally 'out of place' in New Zealand (Milton 2011).
7 One of the earliest Australian children's animal stories to promote the naturalist trope was Edith Pedley's (1997) *Dot and the Kangaroo*, written in 1899 and discussed in detail in A. Taylor (2014).
8 Australians celebrate their legendary bushrangers, such as Ned Kelly and Captain Thunderbolt. They are figures that epitomise the antiauthoritarian spirit of the early convict settler colony. So referring to possums as 'urban bushrangers' is, by association, a term of endearment.
9 I have not witnessed any encounters between possums and children during my research at ANU University Preschool and Child Care Centre, but I have been told many stories about them and seen plenty of evidence of their presence in the playground. I draw upon past possum encounters told to me by Eileen Webster, the centre's assistant director and a staff member for over three decades.

References

Bock, A., 2011. The possum wars. *The Age*, 1 May. Available from: www.theage.com.au/victoria/the-possum-wars-20110430-1e2dt.html [Accessed 10 February 2018].

Bozek, C.K., Prange, S. and Gehrt, S., 2007. The influence of anthropogenic resources on multi-scale habitat selection by raccoons. *Urban Ecosystems*, 10(4), 413–425.

Carter, M., 1992. *Possum goes to school*. Cairns, Australia: Childerset.

Corman, L., 2011. Getting their hands dirty: Raccoons, freegans, and urban 'trash'. *Journal for Critical Animal Studies*, 9(3), 28–61. Available from: http://academicpublishing-platforms.com/downloads/pdfs/jcas/volume1/201112281013_JCAS_vol3_2011_2.pdf [Accessed 10 February 2018].

Couzens, V., 2011. *Kooramook yakeen: Possum dreaming, Aboriginal cultural stories*. Culture Victoria, Victorian Government. Available from: www.cv.vic.gov.au/stories/Aboriginal-culture/possum-skin-cloaks/kooramook-yakeen-possum-dreaming-by-vicki-couzens/ [Accessed 10 February 2018].

Cubit, S., 1998. Building of the fur trade: An introduction to Tasmanian skin sheds and snaring huts. *Historic Environment*, 14(1), 10–18.

Edmonds, P., 2010. Unpacking settler colonialism's urban strategies: Indigenous peoples in Victoria, British Columbia, and the transition to a settler-colonial city. *Urban History Review*, 38(2), 4–20. Available from: www.erudit.org/fr/revues/uhr/2010-v38-n2-uhr3707/039671ar.pdf [Accessed 10 February 2018].

Elliott, J., n.d. *The gift of the day*. First Voices: Sencoten. www.firstvoices.com/en/SENCOTEN/ [Accessed 10 February 2018].

Fox, M. and Vivas, J., 1983. *Possum magic*. Adelaide, Australia: Omnibus Books.

Gavin, P., Kazacos, K. and Schulman, S., 2005. Baylisascariasis. *Clinical Microbiology Reviews*, 18(4), 703–718.

Haraway, D.J., 2008. *When species meet*. Minneapolis: University of Minnesota Press.

————., 2016. *Staying with the trouble: Making kin in the Chthulucene*. Durham: Duke University Press.

Instone, L., 2014. Unruly grasses: Affective attunements in the ecological restoration of urban native grasslands in Australia. *Emotion, Space, and Society*, 10, 79–86.

Jensen, K., 1989. *Possum in the house*. Milwaukee: Gareth Stevens.

Johnson, V., 2007. *Clifford Possum Tjapaltjarri*. Adelaide: Art Gallery of South Australia.

Kerle, A., 2001. *Possums of Australia: The brushtails, ringtails and greater glider*. Sydney: UNSW Press.

Lake-Thom, B., 1997. *Spirits of the earth: A guide to Native North American nature symbols, stories, and ceremonies*. New York: Plume.

Lawrence, D. and Clover, G., 2011. *Sammy Squirrel and Rodney Raccoon to the rescue*. Vancouver: Granville Island Publications.

Lorimer, J., 2007. Nonhuman charisma. *Environment and Planning D: Society and Space*, 25(5), 911–932.

————., 2014. On auks and awkwardness. *Environmental Humanities*, 4, 195–205.

Massey, D., 2005. *For space*. London: SAGE.

McKinney, M.L., 2002. Urbanization, biodiversity, and conservation. *BioScience*, 52, 883–890.

Milton, K., 2011. Possum magic, possum menace: Wildlife control and the demonization of cuteness. *In*: C. Freeman, E. Leane and Y. Watt, eds. *Considering animals: Contemporary studies in human-animal relations*. London: Routledge, 67–79.

Nagy, K. and Johnson, P.D. II, eds., 2013. *Trash animals: How we live with nature's filthy, feral, invasive, and unwanted species*. Minneapolis: University of Minnesota Press.

New South Wales Government Office of Environment and Heritage, 2016. *Brushtailed possum*. Available from: www.environment.nsw.gov.au/animals/TheBrush-tailedPossum.htm [Accessed 10 February 2018].

Nowak, R.M., 1991. *Walker's mammals of the world*. Baltimore: Johns Hopkins University Press.

The Nut Job, 2014. Animated feature film. Directed by P. Lepeniotis. Toronto: ToonBox Entertainment.

O'Sullivan, S., Creed, B. and Gray, J., 2014. 'Low down dirty rat': Popular and moral responses to possums and rats in Melbourne. *Relations Beyond Anthropocentrism*, 2(2), 59–77.

Pacini-Ketchabaw, V. and Nxumalo, F., 2015. Unruly raccoons and troubled educators: Nature/culture divides in a childcare centre. *Environmental Humanities*, 7, 151–168.

Pedley, E.C., 1997. *Dot and the kangaroo*. Sydney: Angus and Robertson. Reproduction of original Angus and Robertson 1906 print edition prepared by Sydney University Library, 1997. Available from: http://adc.library.usyd.edu.au/index.jsp?page=home&database=ozlit [Accessed 9 February 2018].

Pettit, M., 2010. The problem of raccoon intelligence in behaviourist America. *British Journal for the History of Science*, 43(3), 391–421.

Possum wars, 2013. Video documentary. Directed by B. Permezel. Melbourne: 360 Degree Films, Screen Australia and Film Victoria.

Power, E.R., 2009. Border-processes and homemaking: Encounters with possums in Australian suburban homes. *Cultural Geographies*, 16(1), 29–54.

Pratt, M.L., 1992. *Imperial eyes, travel writing and transculturation*. London & New York: Routledge.

Raccoon Nation, 2014. *In*: *The nature of things*. Television program. Canadian Broadcasting Corporation.

Sullivan, R., 2006. *Urban possums*. ABC Science. Available from: www.abc.net.au/science/articles/2006/09/07/2041855.htm [Accessed 10 February 2018].

Taylor, A., 2014. Settler children, kangaroos, and the cultural politics of Australian national belonging. *Global Studies of Childhood*, 4(3), 169–182.

Taylor, H.P., 1996. *Brother Wolf: A Seneca tale*. New York: Farrar, Straus, Giroux.

Zahara, A. and Hird, M., 2015. Raven, dog, human: Inhuman colonialism and unsettling cosmologies. *Environmental Humanities*, 7, 169–190. Available from: http://environmentalhumanities.org/arch/vol7/7.9.pdf [Accessed 10 February 2018].

6 Indigenous child-dog relations

A recuperative ethics of kinship obligation

> All critters share a common 'flesh', laterally, semiotically, and genealogically. Ancestors turn out to be very interesting strangers; kin are unfamiliar (outside what we thought was family or gens), uncanny, haunting, active.
>
> (Haraway 2016, p. 103)

> An ecology of obligations . . . makes us capable of being better obligated to and obligated by other beings, on other trajectories.
>
> (Despret and Meuret 2016, p. 27)

Human-dog kinship

There is ample material and allegorical evidence to suggest that dogs and humans have shared a special relationship at least since the last ice age, making it nigh impossible to think about dogs outside of the context of human culture and to think about human culture without dogs (McHugh 2006, pp. 18–19). Noting that dogs were the only animals living with Inuit and Australian Aboriginal people pre-colonisation, James Serpell (1996) concludes that symbiotic dog-human relations were fundamental to hunter-gatherer societies and served an intermediary function for other human-animal relationships. This special relationship between humans and dogs is registered in the ecological cosmologies of both Inuit and Australian Aboriginal people as a form of kinship, and it also opens up the question of cross-species kinship more generally, as a 'foundational condition of human life' (Rose 2011, p. 4).

It is important to first acknowledge that the consideration of dogs as kin is not exclusive to indigenous societies. As anthropologist Marilyn Strathern (1992, p. 12) points out in her genealogy of middle-class English kinship, the infantilisation and petting of prized animals shares cultural origins with the fussing over or petting of children as unique and special individuals within increasingly small nuclear families. Since the emergence of the anglophone nuclear family, pet dogs, in particular, have taken on the status of child surrogates and siblings. This regard for dogs as family members is self-evident within the classics of anglophone popular culture (from *Lassie Come Home* to *My Dog Skip*), but also has been verified in broad-based attitudinal surveys such as those cited by Gail Melson (2005) in America and Adrian Franklin (2006) in Australia.[1]

Central to the regard for dogs as members of the modern western family is the notion of domestication, which stresses that we have deliberately bred dogs to be docile and obedient to satisfy our human needs and to fit into our human lives. This is a human-centric and assimilatory schema in which humans run the show as the breeders, the masters, and the owners of pet dogs. As part of this assimilatory schema, the degree to which specially bred dogs are regarded as more or less 'cute' is directly related to the tendency to infantilise and/or anthropomorphise them. The sentimental anthropomorphisation of dogs that typifies western notions of dogs as faithful family members is based upon a dualistic schema that situates everything human, or human-made, in the cultural realm and everything else, including other animals, in the natural realm. As we have previously noted, the 'hyperseparation' of humans from the rest of the world is fed by the premise of human exceptionalism, or the belief in humans' exceptional capacity and destiny to dominate, control, manage, exploit, and improve on nature at will (Plumwood 1993). Dogs can only be brought into the human, or 'cultural', side of the divide if they are domesticated enough to fully submit to human mastery and control, to become dependent and child-like, to become surrogate children.

This is a completely different kind of human-dog kinship schema to those of Inuit and Australian Aboriginal people that we consider in this chapter. In fact, we deliberately focus upon Inuit children's relations with qimmiit (sled dogs) in the Canadian Artic and Arrernte and Warlpiri children's relations with camp dogs in the Central Australian desert to help us think beyond western notions of domestication and beyond anthropomorphising and infantilising notions of pet dogs as surrogate child members of their human families. The indigenous accounts and texts that we draw upon are not bound by the sentimental, romantic, and anthropomorphising traditions that pervade western representations of child-dog relations. Instead, they emphasise the pragmatic, the pedagogical, the ecological, the cosmological, and the ontological significance of children and dogs sharing inheritances and growing up well together.

We are particularly interested in the ways that indigenous relational ontologies provide otherwise to western ways of understanding child-dog relationships, because they assume a 'common flesh', as Donna Haraway (2016, p. 103) puts it, 'laterally, semiotically, and genealogically', and because they produce the conditions of possibility for what Vinciane Despret and Michel Meuret (2016, p. 27) refer to as an 'ecology of obligations'. However, we are also acutely aware that there is a direct correspondence between the degrees of colonial dispossession and displacement and the extent of intergenerational transferal and enactment of these indigenous relational ontologies. The contemporary indigenous accounts that we draw upon suggest that the more extreme the decimation of traditional lifestyles and the more broken the relations with the land, the more important children's kinship relations with dogs become to the process of cultural-ecological recuperation. For our purposes, these indigenous accounts of multispecies kinship also serve to counter the powerful normative effects of western discourses that separate humans off from other living beings through the intersecting projects of animal domestication and human mastery and control (Plumwood 1993).

In addition to these Inuit, Arrernte, and Warlpiri accounts of human-dog relations, our thinking about the possibilities for human-animal kinship are informed by human-animal relational philosophies emerging at the interstices of science studies, ethology, and the environmental humanities – most pertinently those of Donna Haraway (2008, 2016), Vinciane Despret and Michel Meuret (2016), Deborah Bird Rose (2011), and Thom van Dooren (2014). With Haraway (2016, 2017), we think beyond the lineages of patriarchal 'natal' kinship to consider all earthlings as 'oddkin' evolving laterally as well as genealogically across entangled networks of diverse biosocial ecologies. With van Dooren (2007), we attend to vertical intergenerational inheritances that facilitate the passing of embodied knowledges from previous to future generations. With Rose (2011), we reflect upon the ethical responsibility to care for multispecies kin in this time of mass extinctions. With Despret and Meuret (2016), we extend these ethical concerns to consider how obligation, as the product of 'cosmoecologies', might point us towards new and recuperative modes of multispecies living in a damaged world.

Although we set up a dialogue between these relational philosophies and the Inuit and Aboriginal accounts of child-dog relations in this chapter, we are cognisant of the fact that they have radically different origins and trajectories. We do not wish to elide these differences, nor to infer that these ways of knowing are reducible to each other. The relational human-animal philosophies have emerged from a critical conversation within their own western knowledge traditions. Although they make a deliberate break with the structuring conventions of western humanist thought, they are nevertheless born of it. The indigenous ontologies shaped (at least partially) by child-dog kinship relations are embodied and locally emplaced knowledges. They emerge from their respective indigenous lands. Despite their different forms and trajectories, we believe that generative possibilities can come from putting these western relational philosophies and indigenous ontologies into conversation with each other (Thomas 2015; van Dooren *et al.* 2016).

As a final caveat, we caution that there are no universals when it comes to children's relations with other species. Any exploration of the significance of children's multispecies relations in their immediate common worlds 'must be undertaken one world at a time' (Taylor 2013b, p. 368). In the following sections, we attend to the ontological and geo-cultural specificities of kin relations between Inuit children and qimmiit, and Arrernte and Warlpiri children and camp dogs, by contemplating the 'the various diversities that they both produce and depend upon' (van Dooren 2007, p. 89). We trace the colonial interventions that have threatened to break these kinship continuities and ties and examine some recent Inuit and Indigenous Australian efforts to recuperate them.

In our concluding reflections on the broader ethical implications of indigenous child-dog kin relations, we review the pedagogical and cosmoecological principles that are illuminated by the indigenous accounts, and make connections between these principles and recuperative eco-philosophies of Haraway and of Despret and Meuret.

Inuit children and qimmiit

Before Inuit were moved into European-style settlements in the mid-20th century, relationships with animals were tantamount, and absolutely fundamental to their nomadic hunting existence in the North American Artic. According to ethnographic accounts, Inuit divide animals into two main groups – those that are hunted and those that hunt (Laugrand and Oosten 2016). Within the latter group, qimmiit, an Arctic dog breed (literally 'many dogs' in Inuktitut), were their close hunting partners. Inuit camped with qimmiit on the tundra in extended family groups, and sled dog teams (*qimmiijaqtauniq*) provided essential mobility across the snow and ice.

Qimmiit were not domesticated 'pets', subject to Inuit mastery and control, but fully fledged agentic partners within the Arctic human-animal hunting assemblage, a critical assemblage for the viability of Inuit and dog lives alike. In fact, relations with all animals, including those that were hunted and killed, were fundamental to survival and subject to strict protocols and rules of respect. They were highly ritualised and based upon notions of human-animal gifting and reciprocity (Laugrand and Oosten 2016). In Inuit cosmologies, all animals were regarded as having a *tarniq*, or soul, and thereby a personhood. However, only qimmiit were regarded as part of the Inuit extended family group and given an *atiq*, or ancestral name (Laugrand and Oosten 2016).

The relationship between Inuit and qimmiit also had central ontological significance. As Frank Tester (2010, p. 134) stresses, it was an affective and formative relationship bound up with the centrally important relations to the land itself: 'How the physical landscape of the Arctic is seen, felt, and experienced was tied to relationships between Inuit and their dogs'. In other words, Inuit relations with qimmiit were integral to Inuit 'ontologies of engagement', to their embodied modes of being in and knowing themselves in the Artic environment (Ingold and Bird-Davis, cited in Laugrand and Oosten 2016, p. 10).

Inuit elders who experienced life on the tundra before European settlement testify that qimmiit were crucial to their livelihoods. The constant challenges of surviving in a frozen land intensified the need for understandings, trust, and respect between Inuit and qimmiit. The stakes were very high. In a recently reprinted interview, originally conducted in 1979, Inuit elder Paulusie Weetaluktuk asserted,

> We needed the dogs to live. If you didn't have dogs you lived in poverty. So you had to respect your dogs. Without dogs you weren't much of a man. . . .
> If he lost his dogs, he lost his ability to travel any distance, or hunt caribou.
>
> (Weetaluktuk 2012/1979)

The bonds of sociality between Inuit and qimmiit were expressed through mutual care and responsibility. Qimmiit cared for their people by protecting them from danger, performing their pack roles in the sled teams and aiding in the hunt and in camp relocation. Inuit cared for their dogs by ensuring they remained strong and healthy: feeding them different kinds of food each season, attending to their paws to avoid ice cuts, and so on. Elder Jimmy 'Flash' Kilabuk (Nowdluk) of Iqaluit explains, 'The dogs were like a member of our team as a family unit as well as our companions. [My] father would treat his dogs like he would treat individuals' (Qikiqtani Inuit Association 2013, p. 15). Yet, like all family relationships, they were not without stresses and tensions. As elders also note, Inuit-qimmiit relationships were sometimes tested by conditions of extreme environmental hardship. In long periods of starvation, Inuit resorted to eating qimmiit, and in stressful times of scarcity, qimmiit could become aggressive to humans outside their Inuit families and to other qimmiit (Qikiqtani Inuit Association 2013; Zahara and Hird 2015).

Qimmiit also played an important role in Inuit spiritual life (Tester 2010). Some Inuit had dogs as helping spirits: 'Punnguq ("spirit dog") could be used to help find animals, to find one's way home in a blizzard, or to help an angukkuq ("shaman") find a kiglurittuq ("bad or terrible spirit") and chase it away' (Tester 2010, p. 134). Qimmiit still feature within Inuit cosmology. A well-known creation story about a marriage between an Inuit woman and male qimmiq accounts for the origins of white people (*qallunaat*):

> A young woman married a qimmiq and gave birth to qimmiit. Because she was poor and could not take care of them, she made a boat for the young dogs, setting up two sticks for masts in the soles of her boots, and sent puppies across the ocean. . . . They arrived in the land beyond the sea and became the ancestors of the Europeans.
>
> (Qikiqtani Inuit Association 2013, p. 15)

The naming of qimmiit, but no other animals, is a key indicator of their unique kinship status in Inuit society. Like newborn babies, qimmiit puppies were given the name or *atiq* of a selected family member, usually deceased. The *atiq* carries with it the social attributes of the namesake. It determines what kinds of personal qualities children and qimmiit will embody. Of additional significance, the *atiq* also designates specific kinship relations and the roles, responsibilities, and obligations that these kinship relations carry. Therefore, 'qimmiit could be fathers, grandfathers, mothers, grandmothers, uncles, aunts, and so forth to their Inuit families' (Qikiqtani Inuit Association 2013, p. 15). Across generations, the assignment of *atiq* to both children and dogs resulted in the intermingling of their genealogies. Qimmiit were given the *atiq* of deceased people and children were given the *atiq* of deceased qimmiit. Qimmiit, as well as Inuit, carried the status of revered ancestors within this cross-generational interspecies kinship system. To put this in van Dooren's (2007) terms, the system of *atiq* bound Inuit children and young dogs together in an entanglement of multispecies intergenerational inheritances.

Puppies lived inside the igloos until they were old enough to survive the cold. At this point, they were taken outside to live with their qimmiit packs. Qimmiit pups and Inuit children grew up in sibling-like relationships, and together they learnt the importance of reciprocal kinship obligations. This process began when a carefully selected qimmiq pup was gifted to a child and through this exchange, both the child and the pup started to 'learn about friendship and familial solidarity' (Tester 2010, p. 134). Indeed, it was through their maturing relationships with qimmiit that children reached adult status. For instance, Zahara and Hird (2015, p. 181) explain, 'Inuit boys [were] considered men only when they [were] able to successfully support a full qimmiit team'. In reflecting upon his childhood, Benjamin Jararuse (1953/2000, pp. 30–31), Kangirsualujjuaq, recalls his special relationship with his qimmiq:

> I was just a small boy then. I enjoyed sliding and all kinds of games. We had dogs, including one puppy that must have been only two feet from nose to tail – that's how small it was. There were six puppies in all, but we were responsible for certain dogs and that puppy was my dog. I had named it Alaku. I was responsible for feeding it and it followed me around. The others were not interested in following me, but Alaku was my companion.

The dog slaughter

The intimate living relationship between qimmiit and Inuit came to a tragic end with the transition from nomadic camp life to permanent urban settlement in the mid-20th century. At first the dogs came with Inuit into the townships. However, in response to (non-Inuit) public health and safety concerns about qimmiit roaming freely on the streets, the Canadian government instructed the Royal Canadian Mounted Police to catch and kill any qimmiit that were not chained up. The results were devastating, leading to the near extermination of sled dogs. The legacy of this ordinance is that there are roughly 300 qimmiit remaining in Canada – a number far below the 20,000 that existed prior to colonisation (Zahara and Hird 2015). The authorised killing of qimmiit is now referred to as the 'Mountie sled dog massacre', or simply 'the dog slaughter', which has become emblematic of the injustices Inuit suffered at the hands of the settler colonisers. The dog slaughter was a central consideration of the Qikiqtani Truth Commission (2007–2012), an independently commissioned inquiry that investigated the impact of the Canadian government's forced assimilation policies upon Inuit (Qikiqtani Inuit Association 2013).

The Commission found that 'the killing of qimmiit was one of the most traumatic elements of the changes that happened as the Canadian government tightened its hold on the everyday life of Inuit' (Qikiqtani Inuit Association 2013, p. 61). Zahara and Hird (2015, p. 180) note that the law prohibiting qimmiit from running freely in communities was 'part of a governance strategy' aimed, not only at getting rid of qimmiit, but also at undermining Inuit autonomy and transforming them into dependent and thus compliant citizens. In the words of some of the elders, the changes precipitated by the dog slaughter were tantamount to a form of 'imprisonment' (Qikiqtani Inuit Association 2013).

Although snowmobiles replaced *qimmiijaqtauniq* as the main means of transport, they could not 'read' the landscape like dogs could (Tester 2010, p. 134), and without the nuanced communications and reciprocal care enabled by the crucial human – dog partnership, the capacity for survival on the ice and snow was severely compromised. As the Qikiqtani Truth Commission notes, the overall effect of the dog slaughter was to deny Inuit-qimmiit kinship relations and to seriously impair all the interspecies ways of knowing and being in the tundra associated with them (Qikiqtani Inuit Association 2013). Many elders now interpret it as an attempt to annihilate not only qimmiit but also Inuit, likening it to a form of genocide (Laugrand and Oosten 2016, p. 151).

By preventing the passing of embodied knowledges from the previous to the future generations of dogs and humans alike, the dog slaughter can be understood, in van Dooren's (2007, p. 82) terms, as the 'unmaking of both lives and the kin relationships from which that life has been both birthed and nourished'. It effectively put everything at stake: the ontologies of entangled and symbiotic human and dog lives, the socialities of human-dog kinship, systems of reciprocal care, obligation, and responsibility, patterns of mutual survival, practices of love, and relationships with time and place. The death of qimmiit was the 'kind of death that undermines the possibility of local adaptation to local conditions' (van Dooren 2007, p. 82). It triggered the process that Deborah Bird Rose (2004) refers to as 'double death' – death that does not regenerate life, but through ignoring multispecies interdependencies and preventing intergenerational inheritances, cascades into a series of compounding and collective extinctions.

Ecological-cultural recuperation

As the snowballing consequences of the catastrophic loss of qimmiit kin becomes more apparent, and in the face of growing awareness about climate change, the recuperation of traditional Inuit knowledges has become an increasingly important project for many Inuit communities (Zahara and Hird 2015). Ice caps melts, precipitation increases, and sea level rises are creating a much more variable and uncertain Arctic, increasing the pressure upon Inuit to try and recover their relations with the land. Any form of recuperation necessarily includes acknowledging the significance of qimmiit (Andre 2007). Specifically, story-telling about Inuit-qimmiit modes of kinship has become a focus for elders wishing to pass on traditional knowledges to children and to mitigate increasingly challenging ecological futures.

An excellent example is the *Kamik* series of children's picture books adapted from the teachings of Inuit elders in northern Canada and situated within a contemporary Nunavut community. The series is published by Inhabit Media Inc., an Inuit company whose mandate is to document and preserve Inuit stories, knowledge, and teachings. The three *Kamik* books recount stories that are based upon elders' recollections of growing up with qimmiit and are told from the perspective of the main character, Jake, an Inuit boy, and Kamik, his sled dog puppy. As Jake and Kamik grow up together, their closely imbricated relationship develops and deepens. It is only in partnership that they can learn a sequence of essential life skills and achieve their coming of age milestones. All this happens under the watchful eye of the elders. Each book focuses upon a maturational milestone and stresses the mutual care and responsibilities that characterise the child-qimmiq relationship.

In the first book, *Kamik, an Inuit Puppy Story* (Uluadluak and Leng 2012), Jake's grandfather (*ataatasiaq*) teaches him how to persevere with the difficult task of training a playful puppy, the importance of choosing the right name (*atiq*) for a sled dog, and the responsibilities and joys of building a close relationship with a qimmiq. *Ataatasiaq* communicates these lessons by telling Jake stories about the olden days when Inuit and qimmiit always lived together. For instance, he describes his own grandmother's skill at training puppies, how she 'raised them in a similar way to raising a child', how she used to speak to the puppies while stretching their muscles, and how she 'used to name [her] puppies after dogs who had passed away' (Uluadluak and Leng 2012, p. 5). He tells Jake how important it is to spend a lot of time with his dog, likening the unfolding child-dog relationship to be 'more like building a good friendship than raising an animal. Eventually, they start to understand you and you start to understand them' (Uluadluak and Leng 2012, p. 8). Jake's grandfather also describes how his dogs once saved his life by bringing him home through a blizzard. At the end of the book, when Jake leaves his grandfather's house, he makes the decision to give Kamik an Inuktitut name. The *atiq* he chooses is Tuhaaji, the name of his grandfather's best lead sled dog.

In *Kamik's First Sled* (Sulurayok and Leng 2015), the second book in the series, Jake's grandmother (*anaanatsiaq*) recalls how children and puppies used to learn how to pull and ride sleds. Gifting Kamik his first backpack and Jake his first sealskin sled, she encourages them to go out of town to practice sledding. The inexperienced pair run into many of the problems that *anaanatsiaq* warns about. Kamik gets distracted by and chases a snow hare, and when a blizzard starts blowing in, Jake worries that he will not be able to see his way back home. Fortunately, Jake remembers the stories of qimmiit smelling their way home, and he leaves it up to Kamik to lead them safely back to his grandmother's house. After they arrive safely, Jake realises that they are both growing up, and he decides it is now time for Kamik to spend the nights outside on the porch. This symbolises a right of passage for Kamik as he transitions from being a puppy living inside to joining the adult dogs living outside.

In the final book, *Kamik Joins the Pack* (Baker and Leng 2016), Jake and Kamik learn the principles of dog sledding from Jake's uncle, who is an expert musher. Uncle reminds Jake about the many skills he will need to learn: how to rebraid and sew ropes and harnesses, how to build sleds and doghouses, how to fix sleds, and how to 'keep dogs healthy so they that don't get hurt when they are running with the pack' (Baker and Leng 2016, p. 8). Jake notices that his dream of 'being a good musher' will take a lot of hard work, and he becomes overwhelmed when he realises his ignorance. But his uncle reassures him by saying, 'You don't need to learn everything all at once. You can learn alongside your dogs' (Baker and Leng 2016, p. 13). The book ends when Jake's uncle offers to put a harness on Kamik and attach him to the end of his team's line. Kamik rises to the challenge and behaves like a smart sled team dog. Jake is now sure that one day, if they both persevere, Kamik will be the best lead dog, like Tuhaaji, his *atiq* namesake.

The elders' stories testify to the ways that Inuit children and dogs grew up together in sibling-like relationships of intimacy, trust, and reciprocity under the guidance of the grandparents and other significant relatives. They reiterate how mutually dependent child-dog relations have been integral to Inuit ontologies and recount the everyday duties of mutual care and love that obligated both children and dogs to persevere with the challenges of learning together how to survive on the tundra. Because most Inuit children no longer have the opportunity to grow up with dogs, stories such as these play a vital symbolic role in cultural recuperation and maintenance. They ensure that Inuit children do not forget that humans always lived with dogs in the Artic and that mutually sustainable life in this challenging environment necessitates partner species care and respect.

We read the messages that elders wish to convey to Inuit children about how they grew up and learnt with qimmiit as a reminder about the significance of reclaiming multispecies kin relationships in ecologically challenging times. Through narrating the significance of lateral Inuit-qimmiit extended kin relations in the past, the Kamik stories also underscore the importance of continuing to transmit intergenerational, or vertical, teachings in the present. In the absence of ongoing embodied knowledge transmission between generations and species, it is only stories such as these that can now 'form part of ongoing duties or practices of respect that are about maintaining animals, traditions, environments, and human community, into the future' (van Dooren 2007, p. 87).

Central Australian Aboriginal children and camp dogs

As with the Inuit in Canada, since colonisation, the majority of Australian Aboriginal peoples have been dispossessed and displaced from their ancestral lands and disconnected from the key multispecies relations pertaining to those lands. Those who still live on their own country tend to be in remote and/or arid areas that are either inaccessible or have less economic value to settler society. In the Central Australian desert region, there are still many different language groups of Aboriginal people who live on their own country, albeit in European-style settlements or 'Aboriginal bush communities'. These include Arrernte, Luritja, and Warlpiri peoples. There is also a significant population of Aboriginal people from all of these same language groups who live in fringe settlements or 'Aboriginal town camps' on the outskirts of Alice Springs, the region's main urban centre. Alice Springs is built upon Mparntwe country, the traditional land of the Central Arrernte people. In all of these bush communities and town camps, people live with dogs.

Prior to colonisation, the Aboriginal people of the Central Australian desert region lived on and moved about their country with dingoes, a canine subspecies that is believed to have migrated down from Asia about 4,000 years ago (Meehan *et al.* 1999). Dingoes were neither bred nor domesticated by Aboriginal people, but were their constant and only nonhuman companions. This relationship was, and in some remote northern communities still is, voluntary and mutually benefi-cial, but dingoes have never been 'domesticated'. They found/find their own food and mates, and they followed/follow their own pack rules of sociality, but they did/or do all of this in close proximity and/or in tandem with humans. In the desert regions, dingoes were close companions and hunting partners and also provided extra warmth on cold winter nights (Wingfield and Austin 2009, p. 14). Since parts of the Central Australia desert have becoming leasehold cattle stations and Alice Springs has been established as a service centre, many working dogs and pet dogs have been brought into the area and have interbred with dingoes. Today, Aboriginal people of this desert region live with an admixture of camp dogs, including introduced breeds and dingo cross-breeds. They do not distinguish din-goes from other dogs – they are all camp dogs.

Dog dreamings

Dogs are not only companion animals – they also have totemic ancestral status. This status is accorded through the dog dreaming stories that belong to specific places and link specific routes, or dreaming tracks, across the country. Dream-ing stories are creation stories that explain how everything came into being. They tell of the journeys and critical events of ancestral beings and feature shape-shifting totemic animals, including dingoes and dogs, who sometimes transform into humans and always into permanent features in the landscape. Dreaming stories are one of the main ways that cultural knowledges are passed down through generations. They are not just stories of the past, but highly adap-tive, present living stories (Turner 2010). To remain alive, they must be passed on to children.

Dreaming stories also explain the significance of particular totemic ancestors to particular groups of people, and thereby determine the order of specific multi-species kinship groups. As Rose (2011, p. 18) explains, 'Multispecies kin groups are the result of creation, and the "Dreaming" applies to the ancestors of these groups'. So, for instance, dog dreaming stories not only explain how the dingo/dog ancestors created certain features of the landscape, but also designate which people are related to these totemic dingo/dog ancestors and thereby have a spe-cial kinship relationship with living dogs. Rose (2000, 2011) has written exten-sively about the significance of ancestral dog dreaming stories in the Victoria River region in the Top End of the Northern Territory. She first learnt about them from Old Tim Yilngayarri, a renowned Yarralin 'dog man' elder with a legendary 'passionate attachment to dogs' (Rose 2011, p. 7). As Rose (2011, p. 23) explains, Old Tim's totemic animal kin were dingoes, but he was 'deeply committed to all dogs'. She speculates that on top of his kinship ties, Old Tim's commitment may have been fortified by earlier district dog massacres carried out by the police. According to Rose (2011, p. 7), Old Tim insisted that *all* humans have dingo/dog ancestors, not just 'dog people' like him – Aboriginal people have dingo ancestry and white people are the descendants of white dogs. In Old Tim's cosmology, 'dingo makes us [all] human' (Rose 2000).

While not all Aboriginal people share these beliefs, there are numerous dog dreamings across the country that account for the physical and spiritual characteristics of particular places and have special significance for the local 'dog people', as they determine their networks of kinship responsibility and obligation. There are two such dog dreamings that we know of in the Central Australian desert. The first narrates the journey taken by ancestral dogs hundreds of kilometres across (mostly) Warlpiri country, from Rabbit Flat in the far central west of the Northern Territory to Ali Curung, a few hundred kilometres north of Alice Springs (*Ali Curung Dog Dreaming* 2012). Many Warlpiri places along this route are associated with the dog dreaming, but Ali Curung (literally 'dog place' in the local language) is its epicentre.

The second dog dreaming is the Central Arrernte 'Wild Dog Dreaming' story of Mparntwe country, Alice Springs. This dreaming story is about a territorial encounter between a local dog and an intruder. In one version of the story (recorded at the 1997 Native Title Tribunal), the local dog was originally a Warlpiri dingo from Nyirripi who 'bred up here' [in Alice Springs] (*Mparntwe Sacred Sites: Wild Dog Dreaming* 2004), suggesting the ways in which Warlpiri and Arrernte dog dreaming stories are interconnected. In 1988, a senior Central Arrernte man and Mparntwe custodian, Basil Stevens, narrated a detailed version of the 'Wild Dog Dreaming' (*Ayeye Akngwelye Mpartnwe-arenye*) story for the children of Mparntwe. His audio-recorded narration was made into an Arrernte reading book for the children of Yipirinya School, an independent Aboriginal school originally set up for children living in the Alice Springs town camps.[2] As well as providing these children with the requisite skills for living in mainstream Australian society, the school's original charter was to keep their culture alive by also educating them about their cultural inheritances, including their dreaming stories, and to do so in their own languages (Taylor 2013a).[3]

To paraphrase Stevens's account, the local dog, Akgwelye, was living happily with his mate and puppies until their life was rudely interrupted by an intruder 'devil dog' that came into their territory from the south. The intruder dog climbed Mount Gillen (Alhekulyele) (an imposing section of the MacDonnell Range escarpment that flanks Alice Springs), where he attacked a girl dog and left her to die. He then threatened to kill Akngwelye's mate and puppies. Akngwelye fought with him. He ripped open the devil dog's stomach, and his guts spilled onto the ground. These entrails became a rocky ridge (Yarrentye) to the west of the township. Both dogs survived the fight, but the defeated intruder limped back through the Gap (Ntaripe, a break in the MacDonnell Ranges immediately south of Alice Springs) and retreated. Also injured, Akngwelye lay down under a sheltering tree. From there he continued to watch over and guard his country. He never returned to his mate and pups, turning into a boulder, which is still to be seen on the northwest edge of the town's central business district. Today, this boulder is a designated sacred site known as Akngwelye Thirrewe and is fenced off with a chain.[4]

The children's 'Wild Dog Dreaming' (*Ayeye Akngwelye Mpartnwe-arenye*) book has played a dual function. It was originally intended to be used as an Arrernte reader within Yipirinya School's vernacular literacy program. Perhaps even more importantly, though, it was written to ensure that this important local dreaming story could be passed on to future generations of Aboriginal children in Alice Springs, many of whom have an extended kinship connection to dogs. It still remains a significant material cultural archive. Basil Stevens's version of the 'Wild Dog Dreaming' story was reenacted at the 2016 National Aboriginal and Islander Day of Celebration (NAIDOC) event in Alice Springs in order to register its pertinence for local Aboriginal children and to keep it alive. It was narrated by local elder and custodian Doris Stuart and dramatised by her grandchildren and great-grandchildren (Finnane 2016).

Inherited responsibilities and obligations

In Alice Springs, Aboriginal intergenerational and cross-kinship species responsibilities and obligations are not only still relevant, but are also operative, despite (or perhaps because of) the lack of respect on the part of many local non-indigenous authorities for Aboriginal dreaming stories and their sacred sites. Doris Stuart told the NAIDOC gathering that when she was a child, the old people told her to stroke the dog rock (Akngwelye Thirrewe) whenever she passed it, as a sign of respect. She added,

> It is my job to protect these places . . . I inherited this responsibility from my Father's Father. It is a hard job, you can see from these photos that a lot of damage has been and continues to be done to our sacred places.
>
> (Cited in Finnane 2016)

This duty and obligation to protect sacred sites associated with dog dreamings also extends, in some cases, to protecting living dogs. For instance, the old people of Ali Curung (literally 'dog country') are guardians of the dogs that live in this Aboriginal bush community. It is also their responsibility to hand their knowledges on to children. The elders teach that some dogs have special powers, and because of this, they must be treated with the utmost respect (Arlpwe Art and Culture Centre 2015). In a recent film, *Staying Safe around Dogs* (Animal Management in Rural and Remote Indigenous Communities 2016), members of the community explain that because Ali Curung is a dog dreaming place, 'everyone has to look after any dogs, you know? They can't hurt them, or they can't tease them, they have to look after them'. Anyone who harms or kills a dog risks dire consequences; they might get sick and 'end up in hospital'. 'Even if we bump a dog by accident [in the car], we have to get out and say sorry to that dog – or else the elders will growl at us'.

The truth is that because of the ravages of colonisation, not everyone does look after dogs. There are plenty of sick, starving, and generally out-of-sorts camp dogs with a reputation for causing trouble wandering around in Aboriginal communities. Uncared-for dogs can also carry scabies and intestinal parasites that pose a health risk for children (Central Desert Regional Council 2014). Camp dogs that roam in threatening packs are commonly referred to as 'cheeky dogs'.[5] There are a number of dog management programs in the Northern Territory primarily aimed at tackling the health and safety concerns related to 'cheeky dogs' in Aboriginal communities. They focus upon population control through free vet sterilisation and treatment programs, but also produce school and community education programs for Aboriginal people. *Staying Safe Around Dogs* is an example of a recent educational resource that encourages children and other community members to 'teach our dogs the right way' so that they don't 'become cheeky and bite you' (Animal Management in Rural and Remote Indigenous Communities 2016).

Growing up 'the right way'

The slogan 'teach our dogs the right way' echoes a common reference to teaching children 'the right way', a mode of informal cross-generational education that is primarily the responsibility of grandparents.[6] It involves ensuring that young people grow up knowing as much as possible about their country, the plants and animals that are part of it, and the kinship connections and obligations associated with its dreaming stories. In her book *Iwenhe Tyerrtye – What It Means to be an Aboriginal Person*, written for her grandchildren and other Aboriginal children in Central Australia, Eastern Arrernte elder M.K. Turner (2010) stresses that all living things are interconnected in visceral and embodied ways through the land: 'We're part of the animals that live there, and birds. . . . We're part of the water there, we're part of the grasses, the medicines, the fruits' (Turner 2010, p. 114). These inter-corporeal kin relations are manifest by practices of eating and being eaten:

> The bush meat comes from the Land, and that was our meat [*kere*] and that *kere* is part of us. . . . We really relate to all those animals because they're part of us. . . . The eaters get eaten. That eats this and you eat that.
>
> (Turner 2010, pp. 165–166)

All animals have dreamings, even those that are eaten. They all need to be treated and talked about 'in the right way . . . then they're satisfied' (Wingfield and Austin 2009, p. 14).

Rose (2011) likens the kinship responsibilities and obligations that Aboriginal people have to the land, or country, and to other animals (and which grandparents teach children) as part of a moral code. She explains that this moral code 'is not so much a rule as a statement of how life works: a country and its living beings take care of their own. Care for country is a matter of both self-interest and interest for others' (Rose 2011, p. 18). It is pragmatic. It ensures the ongoingness of life: 'Country is the living context in which past, present, and future are part of cross-species relationships of care' (Rose 2011, p. 27).

The everyday practice of cross-species kin relations of care, responsibility, and obligation is part of growing up the 'the right way'. There is nothing extraordinary about it, in fact it is quite mundane, but it is nevertheless otherwise to western ways of doing family relations. A sense of 'how life works' (Rose 2011, p. 18) according to the logics of a multispecies and intergenerational ethics of care that are practiced in uneventful, everyday ways can be gleaned from the observational documentary film *Wirriya: Small Boy* (2004), made by Aboriginal filmmaker Beck Cole.

The film follows the daily life of 7-year-old Ricco Japaljarri Martin, a Warlpiri boy originally from the Tennant Creek area (associated with dog dreaming), who was removed from a violent family situation and is now living with a foster family in Hidden Valley, one of the Aboriginal town camps located on the fringes of Alice Springs. It provides a snapshot into the daily routines of a modest, shared life in which Ricco and his new extended family live according to the 'right ways' of intergenerational and cross-species 'relationships of care' (Rose 2011, p. 27).

At the beginning of the film, Ricco introduces himself and the other members of his family. He starts with his dogs, introducing them by name: 'I have nine dogs. Their names are Michael 1, Michael 2, Cindy, Wink, Ringo, Blacky, Rocky, Tommy, and one named Lu-Lu'. He then introduces his foster grandmother, whom he calls 'Nanna Maudie', and finally two of his three foster sisters. He clearly enjoys his family: 'They're helpful. They share, you know. Share a feed', he affirms. Nanna Maudie reconfirms the conviviality of this non-biological extended family: 'We're just all mixed up tribe living here. We just live together, friendly way'.

Ricco's dogs are central to his conceptualisation of a 'helpful' family. He is clearly proud of them and of the fact that he puts a lot of effort into looking after them well – feeding them daily, making sure they always have fresh water, and taking them on lots of walks. Ricco's relationship with his dogs is more than an affectionate and companionable one. He is keen to stress that he and the dogs already hunt for lizards in the hills surrounding Hidden Valley town camp on these walks. Ricco is looking forward to the day when he is old enough to go kangaroo hunting with his dogs. He aspires to be a top hunter and 'gun shooter', like his uncle. He knows that good hunters and good hunting dogs are held in high esteem because they feed the community.

Ricco's life is far from ideal. Like all Aboriginal children growing up in the fringe camps of Alice Springs and on the margins of mainstream Australian society, he faces innumerable challenges (see Taylor 2013a). But we are struck by how his new life with Nanna Maudie and his nine dogs offers him a renewed opportunity for co-shaping relations of responsibility and obligation, for fulfilling his dreaming inheritances, and for growing up 'the right way'. In a culturally and ecologically colonised and damaged world, these non-natal kin relations stitch him back into modes of obligatory intergenerational cross-species care that sustained his country and ancestors for tens of thousands of years.

Conclusion: towards a recuperative ethics of kinship obligation

The indigenous accounts of child-dog kinship in Inuit and Central Australian Aboriginal communities we have provided earlier offer us a glimpse into otherwise modes of understanding what it means to share lives with other species.

They show us that the specialness of child-dog companion relations need not be based upon anthropomorphic sentiments, nor confined to the domestic sphere of the modern western nuclear family. Instead, they reveal that these relations have much broader pedagogical, ecological, and cosmological significance. The broader significance of these relations highlights some guiding ethical principles for composing worlds differently and provides hope that 'maybe, just maybe, other kinds of communities might still be possible' (van Dooren and Rose 2016, p. 79).

Let us reiterate that we are definitely not suggesting that all children should understand their own companion species relations in exactly the same ways that Inuit understand children's familial relations with qimmiit and that the Arrernte and Warlpiri people understand children's kinship with dogs. This would be not only appropriative but also impossible. These Inuit, Arrernte, and Warlpiri forms of child-dog kin relations are geo-ontologically specific and based upon the inherited and embodied knowledges of millennia of cohabitation in their respective tundra and desert lands. They cannot be carbon copied, lifted out of context, and reapplied elsewhere.

However, there are some principles we can glean from the pedagogical and cosmoecological significance of child-dog kinship in these indigenous accounts that help to illuminate the ethical possibilities for recuperating and recomposing multitudes of other local (but ultimately interconnected) damaged common world ecologies. For the remainder of this chapter, we intend to hone in upon these principles and to link them to some of the key contemporary thinking about relational ethics that is emerging from the field of multispecies studies (van Dooren *et al.* 2016). More specifically, through exploring these links and moving to and fro between the indigenous accounts and multispecies scholarship, we will assemble the details and prospects of a recuperative ethics of kinship obligation.

One of the things that strikes us about the Inuit, Arrernte, and Warlpiri elders' accounts is the central pedagogical significance they place on child-dog kin relations. They all make it crystal clear that these relations are part and parcel of the everyday business of teaching and learning about cultural inheritance, and of teaching and learning about how to grow up well together in order to ensure collective well-being. These pedagogies have a pragmatic ethical foundation. The elders stress the intergenerational and coeval responsibilities and obligations that flow from simply being kin and needing to care for each other. Enacting these responsibilities is an integral part of teaching and learning about what it means to be Inuit/Arrernte/Warlpiri. For the children, the pedagogical task is to learn by doing – to fulfil their everyday responsibilities and obligations by learning how to care for their dog kin and, along with them, to learn practical life skills. The elders' pedagogical responsibility is to ensure that they pass dog kinship stories down the generations in order to keep culture alive. In the face of colonisation's cumulative cultural and ecological devastations, all of these pedagogical tasks have become increasingly recuperative.

From her radically different perspective as a North American white-settler descendant and a feminist science studies scholar, Donna Haraway (2008, 2016) also muses upon the pedagogical and ethical significance of kin relations between species. Having reminded us that 'all earthlings are kin in the deepest sense' (Haraway 2016, p. 103), she emphasises the importance of learning to inherit the legacies of human-nonhuman entangled lives and trajectories – learning how to take responsibility for the cascading ecological damage precipitated by colonial/capitalist exploitation. For Haraway, the pedagogical task of learning how to inherit always raises the ethical question of accountability, of what is to be done. She insists that the key to this learning is to 'stay with the trouble' of these messy inheritances, to resist the temptation to reach for yet another heroic and grandiose human-centric fix, and to redouble our efforts to 'make kin' with non-natal and nonhuman others (Haraway 2016). She emphasises that it is this sense of kinship with other earthlings that obliges us to take responsibility and motivates us to think and act collectively. Haraway's vision is modest and hopeful. It is to be enacted in the ordinary, everyday doings of our multispecies relations. By making kin (or oddkin, as she also calls them) and learning *with* them about how to reshape more liveable worlds together, Haraway maintains that we can all contribute the collective reparative task of recomposing damaged lifeworlds (Haraway 2016).

Returning again to the illustrative principles of the indigenous accounts, we note that the pedagogical and ethical imperatives of Inuit/Arrernte/Warlpiri child-dog kin relations are underpinned by cosmoecological premises. They build upon foundational beliefs in common ancestry, whereby all life comes from the land, is interconnected, and must be respected. They recognise child-dog kinship as predetermined, relayed through the creation stories (or the dreamings in Aboriginal Australian communities), and in the Arrernte accounts are materialised through physical sites in the landscape. Within these indigenous cosmoecologies, all creatures are indebted to their (mutual) ancestors for the gift of life that is passed down to them, but they are also indebted to the other creatures with whom they share their lives – whether as prey or as hunters. This is because these coeval relations are symbiotic – they sustain and feed each other. A key principle that infuses these accounts is that the human-dog bonds and partnerships established in the early years of life support an unfolding cosmoecological understanding of the interdependencies and mutual obligations of all life-giving and life-receiving relations, an experiential understanding that is further reinforced by the stories the elders tell.

Speaking about cosmoecological ethical concerns from a very different geocultural temporality and space, continental philosophers/ethologists Vinciane Despret and Michel Meuret (2016) reflect upon new shepherding practices in the south of France, a way of exploring the question of what makes us capable of a mode of obligation that extends beyond our own species (Despret and Meuret 2016, p. 27). They explain how the new connections that emerged over time between the inexperienced shepherds and the previously enclosed sheep prompted them to speculate that new modes of interspecies connection can also establish new 'ecologies of obligation' (Despret and Meuret 2016, pp. 27–28). They stress that these ecologies of obligation are cosmoecological – they are 'ethical . . . relations to the world' (Despret and Meuret 2016, p. 28). They are not just individual obligations of care between the individual humans and nonhuman animals involved. They are cosmoecological because they are also mutually obligated by the need to find new ways of cohabitating in an altered and damaged world, mutually obligated to compose new ways of living together that are ecologically rehabilitating and recuperative (Despret and Meuret 2016, p. 34).

There is no sense of pedagogical inheritance or the intergenerational transmission of ancient geo-ontological knowledges in Despret and Meuret's philosophical reflections. There is scant reference to more-than-human kinship and no reference at all to the significance of child–animal animals. The ethics that emerges from their exploration of an ecology of obligation is based upon reflections of emergent and adaptive modes of interspecies reparative relations. What the Canadian and Australian indigenous accounts hold in common with those of the European ethologists is an acute awareness of the need for recuperation in a damaged world – a form of recuperation that is based upon a cosmoecological ethics of obligation and enacted in the ordinary, everyday practice of interspecies 'earthly companionship' (Despret and Meuret 2016, p. 32). They share the common purpose of seeking 'ways of inhabiting a world that is being destroyed while resisting, locally and actively, this destruction' (Despret and Meuret 2016, p. 30).

A recuperative ethics of kinship obligation has nothing to do with individual (human) altruism or largesse. It has nothing to do with our anthropocentric sentiments for our closest furry companion species. It is about learning how to survive the current ecological train-wreck together, or not at all, because we are already mortal kin, obligated and beholden to each other for the sustainability of our lives. As the indigenous elders know so well, and the key pedagogical and cosmoecological principles of Inuit, Arrernte, and Warlpiri child-dog kin relations readily illustrate, we cannot do it alone. Maybe, just maybe, if we can all find ways of enacting such an ethics within our own common world relations, we might be able to collectively recompose more liveable worlds.

Notes

1 For a more detailed discussion of the treatment of pet animals as children and the treatment of children as pets, see Taylor (2013b, pp. 84–85).
2 The children's Arrernte reader *Ayeye Akngwelye Mpartnwe-arenye* was narrated by Basil Stevens, transcribed by Nanette Sharpe, and illustrated by Thomas Stevens.
3 I (Affrica) was a teacher at Yipirinya School in the late 1980s when *Ayeye Akngwelye Mpartnwe-arenye* was written. For more detailed reflections upon my time at the school, historical information about the Alice Springs town camps, and discussions about Yipirinya children's relations with their animal dreamings, see my article 'Caterpillar Childhoods: Engaging with the Otherwise Worlds of Central Australian Aboriginal Children' (Taylor 2013a).
4 For another paraphrasing of Basil Stevens's 'Wild Dog Dreaming' story, see the newspaper article by Keiran Finnane (2016).
5 The term 'cheeky dogs' has more recently had a shift in meaning, as the popular and humourous 'cheeky dog' drawings of young Warlpiri artist Dion Beasley (Bell and Beasley 2013) have turned it into an endearing form of reference to all Aboriginal camp dogs.

6 In most Aboriginal cultures, references to grandmothers and grandfathers are quite loose and liberal. They are commonly used to denote a second-generation relationship that is not necessarily biological. Children and grandparents have a special relationship, which is marked by respect (on the part of children) and responsibilities (on the part of grandparents).

References

Ali Curung dog dreaming, 2012. Short film. Directed by S. Raabe. Alice Springs, Australia: Indigenous Community Television. Available from: www.youtube.com/watch?v=AQb9MoQwiAQ [Accessed 16 February 2018].

Andre, E.K., 2007. Dogsledding with the Inuit in a warming Arctic. *Phi Kappa Phi Forum*, 87(4), 2–7.

Animal Management in Rural and Remote Indigenous Communities, 2016. *Staying safe around dogs*. Available from: www.amrric.org/news/staying-safe-around-dogs-video-now-online [Accessed 16 February 2018].

Arlpwe Art and Culture Centre, 2015. *Ali Curung dog dreaming: Community*. Available from: www.australianAboriginalartgallery.com.au/community.html [Accessed 16 February 2018].

Baker, D. and Leng, Q., 2016. *Kamik joins the pack*. Iqaluit, Nunavut: Inhabit Media, Inc.

Bell, J. and Beasley, D., 2013. *Too many cheeky dogs*. Sydney: Allen & Unwin.

Central Desert Regional Council, 2014. *Animal management plan*. Available from: www.centraldesert.nt.gov.au/sites/centraldesert.nt.gov.au/files/attachments/2014_animal_management_plan_2014_final_2feb14.pdf [Accessed 16 February 2018].

Despret, V. and Meuret, M., 2016. Cosmoecological sheep and the arts of living on a damaged planet. *Environmental Humanities*, 8(1), 24–36. doi:10.1215/22011919-3527704.

Finnane, K., 2016. NAIDOC celebrates the wild dog story of Alice Springs. *Alice Springs News Online*. Available from: www.alicespringsnews.com.au/2016/07/09/naidoc-celebrates-the-wild-dog-story-of-alice-springs/ [Accessed 16 February 2018].

Franklin, A., 2006. *Animal nation: The true story of animals and Australia*. Sydney: UNSW Press.

Haraway, D.J., 2008. *When species meet*. Minneapolis: University of Minnesota Press.

———., 2016. *Staying with the trouble: Making kin in the Chthulucene*. Durham: Duke University Press.

———., 2017. *Making oddkin: Story telling for earthly survival*. Public Lecture, Yale University, October 23. Available from: www.youtube.com/watch?v=z-iEnSztKu8 [Accessed 9 February 2018].

Jararuse, B., 1953. Alaku, my puppy. *Tumivut: The Cultural Magazine of the Nunavik Inuit*, 12 (special issue 'Qimmiit, Eskimo Dogs'). Reprinted by Nunavik Publications, 2000.

Laugrand, F. and Oosten, J., 2016. *Hunters, predators and prey: Inuit perceptions of animals*. New York: Berghahn Books.

McHugh, S., 2006. *Dog*. London: Reaktion Books.

Meehan, B., Jones, R. and Vincent, A., 1999. Gulu-kula: Dogs in Anbarra society, Arnhem land. *Aboriginal History*, 23, 83–105.

Melson, G.F., 2005. *Why the wild things are: Animals in the lives of children*. Cambridge, MA: Harvard University Press.

Mparntwe sacred sites: Wild dog dreaming, 2004. Documentary. Directed by D. Maclean. Australian Screen, CAAMA Productions. Available from: http://aso.gov.au/titles/documentaries/mparntwe-sacred-sites/clip3/ [Accessed 16 February 2018].

Plumwood, V., 1993. *Feminism and the mastery of nature*. New York: Routledge.

Qikiqtani Inuit Association, 2013. *QTC final report: Achieving saimaqatigiingniq*. Iqaluit, Nunavut: Inhabit Media, Inc. Available from: http://qtcommission.ca/sites/default/files/public/thematic_reports/thematic_reports_english_final_report.pdf [Accessed 16 February 2018].

Rose, D.B., 2000. *Dingo makes us human: Life and land in Australian Aboriginal culture*. Cambridge: Cambridge University Press.

———., 2004. *Reports from a wild country: Ethics for decolonisation*. Sydney: UNSW Press.

———., 2011. *Wild dog dreaming: Love and extinction*. Charlottesville: University of Virginia Press.

Serpell, J., 1996. *In the company of animals: A study of human – Animal relationships*. Rev. ed. Cambridge: Cambridge University Press.

Strathern, M., 1992. *After nature: English kinship in the late twentieth century*. Cambridge: Cambridge University Press.

Sulurayok, M. and Leng, Q., 2015. *Kamik's first sled*. Iqaluit, Nunavut: Inhabit Media, Inc.

Taylor, A., 2013a. Caterpillar childhoods. Engaging with the otherwise worlds of Central Australian Aboriginal children. *Global Studies of Childhood*, 3(4), 366–379.

———., 2013b. *Reconfiguring the natures of childhood*. London: Routledge.

Tester, F.J., 2010. Mad dogs and (mostly) Englishmen: Colonial relations, commodities, and the fate of Inuit sled dogs. *Études/Inuit/Studies*, 34(2), 129–147.

Thomas, A.C., 2015. Indigenous more-than-humanisms: Relational ethics with the Hurunui River in Aotearoa New Zealand. *Social and Cultural Geography*, 16(8), 974–990.

Turner, M.K., 2010. *Iwenhe tyerrtye: What it means to be an Aboriginal person*. Alice Springs, Australia: IAD Press.

Uluadluak, D. and Leng, Q., 2012. *Kamik, an Inuit puppy story*. Iqaluit, Nunavut: Inhabit Media, Inc.

van Dooren, T., 2007. Terminated seed: Death, proprietary kinship, and the production of (bio)wealth. *Science as Culture*, 16(1), 71–93.

———., 2014. *Flight ways: Life and loss at the edge of extinction*. New York: Columbia University Press.

van Dooren, T., Kirksey, E. and Munster, U., 2016. Multispecies studies: Cultivating arts of attentiveness. *Environmental Humanities*, 8(1), 1–23.

van Dooren, T. and Rose, D.B., 2016. Lively ethnography: Storying animist worlds. *Environmental Humanities*, 8(1), 77–94.

Weetaluktuk, P., 2012. Healthy diet for dogs. 1979 interview reproduced in *The Fan Hitch: Journal of the Inuit Sled Dog International*, 14(3). Available from: http://thefanhitch.org/V14N3/V14,N3Tumivut.html [Accessed 16 February 2018].

Wingfield, E.W. and Austin, E.M., 2009. *Living alongside the animals: Anangu way*. Alice Springs, Australia: IAD Press.

Wirriya: Small boy, 2004. Documentary film. Directed by B. Cole. Alice Springs, Australia: Australian Screen, CAAMA Productions.

Zahara, A. and Hird, M., 2015. Raven, dog, human: Inhuman colonialism and unsettling cosmologies. *Environmental Humanities*, 7, 169–190. Available from: http://environmentalhumanities.org/arch/vol7/7.9.pdf [Accessed 10 February 2018].

Conclusion
Relational ethics for entangled lives

Time for a paradigm shift

We introduced this book by raising two questions. Who and what constitute the common worlds that children and animals co-inherit and co-inhabit? And, how might we approach these common worlds in ways that attend to the common good of all its constituents? The first is an onto-epistemological question, the second an ethical one. Both are compositional and inclusive. Both hint at the question of agency.

From the start, we have asserted our belief that neither of these questions can be adequately answered without making a paradigm shift beyond humanism's singular focus upon the social and beyond its foundational premise of agency as an exclusively rational, autonomous, individual, human capacity. Accordingly, throughout the book, we have moved from a number of counter premises. We have insisted that children grow up in more-than-human worlds, not just in human societies. Within these common worlds, their lives, fates, and futures are bound up with those of other animals. We have emphasised the productive nature of relations, including child–animal relations. This infers a relational notion of agency. It means understanding that no being acts alone. It is *relations* – not self-contained entities – that make things happen. It is relations that shape and reshape worlds.

Another of our baseline premises has been that the common worlds that children and animals inherit are not only damaged, but increasingly ecologically precarious. This is founded upon the scientific evidence of a series of accelerating and anthropogenic bio-geospheric system change indicators, including mass species extinctions and global warming (Steffen *et al.* 2007). We have made direct connections between the anthropogenic ecological damage on the local grounds on which children and animals live, and the historical trajectories of the key globalising human projects – colonisation, industrialisation, and modernisation. We have also noted that it is the myopic, anthropocentric humanist belief in the inevitability of human progress, development, and expansion that has driven these grandiose and ecologically disastrous projects.

In Chapter 1, we reiterated (following Haraway 2016, p. 12) that the ideas we use to think with matter. By mattering, we were referring to the ideas that we value most, as well as quite literally, how ideas materialise. The dire ecological consequences of deluding ourselves that humans are exceptional, and therefore able to act upon an exteriorised environment with impunity to serve our own interests, provide the most prescient reminder that the paradigms we use to think with matter. As Anna Tsing (2012, p. 506) succinctly puts it, 'Conceptualising the world and making the world are wrapped up with each other'. Understanding the formative relationship between paradigms or knowledge traditions, and modes of being and acting in the world, not only alerts us to the profound dangers of human-centric exceptionalist conceits, but it also re-opens the possibility for at least some modest ecological recuperation. If the way we think is bound up with the ways we act in relation to others, and in turn with the ways in which worlds are made, it must be possible to contribute to recuperating and reshaping more liveable common worlds for all the constituents, through adopting (more-than-human) collective and inclusive modes of thought and action.

We remain hopeful that change is still possible, but only if we can move beyond the solipsisms of individual humanist thought and can think and act collectively and inclusively with others in our common worlds. This hope has been reinforced by our engagement with non-divisive and hence 'otherwise' ontologies (Povinelli 2016). Firstly, as we have noted, not all human knowledge traditions move from the premise that humans are separate and exceptional, and not all humans are equally responsible for the ecological damage that flows from anthropocentric thought and action. In Chapter 6, we drew attention to some of non-divisive onto-epistemologies of indigenous peoples in Canada and Australia, which have endured (in varying degrees) the ravages of colonisation of the mind as well as the land. Secondly, across the years of conducting ethnographic research with children and animals, we have consistently observed that, prior to formal schooling, young children are more likely to recognise the subjectivity of other creatures and to make connections *with* them in embodied and multisensory ways. They are less inclined to separate themselves off, objectify other beings, and know *about* them from a distance.

Throughout this book, we have responded to the declaration of the Anthropocene (and the debates that it has triggered) as a wake-up call requiring new paradigms for reconsidering our place and agency in the world. Knowing that there are already other ways of knowing and acting *with* the world (rather than only ever about and upon it) reconfirms our belief that such a paradigm shift is possible. And, finally, our grounded observations of child–animal encounters (featured in Chapters 2, 3, and 5) have offered us a window into the *how* of minor interspecies achievements and the relational ethical possibilities that they portend.

Interspecies relational ethics

It has not been our intention to prescribe a universal ethical code – to deliver a one-size-fits-all set of moral imperatives for child–animal relations in the face of the Anthropocene. Instead, we have endeavoured to tease out the ethical affordances of children's and animals' materially and semiotically entangled lives on the local grounds of their common worlds – paying close attention to the specificities of the geo-historical-ecological convergences, complexities, and challenges at each locale. It is on these local grounds – with all their particularities and idio-syncrasies – that we have identified kernels of ethical possibility in the unique ways that child–animal relations unfold. This has included the ethical possibilities in the face-to-face child-wildlife encounters we witnessed in and around the early childhood education centres, as well as the ethical possibilities of particular children's engagements with particular animal story traditions and genres.

Our thinking about the ethical possibilities of these child–animal relations has been supported by the ideational 'gifts' of the many fine multispecies and environmental theorists and philosophers that we acknowledged in Chapter 1 (for instance, Anderson 1995; Gibson *et al.* 2015; Haraway 2008; Lorimer 2015; Povinelli 2016; Rose *et al.* 2012; Stengers 2016; Tsing 2015). It has been even more directly informed by the specific ethical propositions with which we have engaged in Chapters 2 to 6 (Despret and Meuret 2016; Haraway 2016; Hird 2010, 2012, 2013; Puig de la Bellacasa 2010; Rose 2004, 2011; van Dooren 2014; van Dooren and Rose 2012). These varied (but nonetheless all relational) propositions helped us to reflect upon the range of child–animal entanglements unfolding on local Canadian and Australian common grounds. They helped us to clarify exactly what kinds of interspecies relational ethical possibilities we were witnessing.

Across all our considerations of the ethical dimensions of child–animal rela-tions, particularly in our respective settler colonial contexts, we owe a particular debt to Deborah Bird Rose (2004, 2011, 2012). Rose has consistently pointed out that ethical responses are enabled when we pay attention to the world around us, are drawn into relationship with earth others and recognise that our lives are already entangled, in all sorts of temporal, synchronistic, and mortal ways. In the preface to her blog – 'Love and Extinction' – Rose declares that her life's purpose 'is to explore the entangled ethics of love, contingency, and desire in the face of almost incomprehensible loss'. She also acknowledges that the time she spent liv-ing and working closely with Indigenous Australians has helped her to realise and do this (Rose 2016). Her extensive and probing thinking about the possibilities for interspecies relational ethics, in the settler-colonised lands and in the face of anthropogenic extinctions has had a profound influence upon our own.

In spite of the fact that we have both undertaken studies of child–animal rela-tions in settler-colonised lands, each specific example of child–animal relations we discuss in the book has its own set of contingencies and conditions of possibil-ity. We have paired up examples from Canada and Australia because together they throw light on a particular form of interspecies relational ethics. However, these paired examples are not exactly the same. As we have often previously stated, this is because the specificities of relations matter, and because things turn out differ-ently in different places.

It matters which animals draw which children into relationship. For instance, the deer and kangaroos that feature in Chapter 2 and the raccoons and possums that feature in Chapter 5 are all 'charismatic' wild animals (Lorimer 2010). They are mammals with large eyes and faces that are relatively easy for us humans to identify with and thus be attracted to. For the same reasons, they are all animals that have been sentimentally and anthropomorphically characterised in children's

literature and popular culture, making it likely that the children in our ethnographic studies were already familiar with them and felt some form of affection
towards them, even before the face-to-face encounters that we witnessed. No
doubt it was the shy charisma of the deer and kangaroos that pre-disposed the
children to extend an 'ethics of multispecies conviviality' (van Dooren and Rose
2012) to these urban dwellers, and the 'cute' charisma of the otherwise pesky raccoons and possums that encouraged the children (and educators) to persist with
an ethics of 'staying with the trouble' (Haraway 2016) when residing with them.
Regardless of their common charismatic appeal, however, deer and kangaroos
are not the same animals. Neither are possums and raccoons. They do not live in
the same places, and they are positioned very differently within their respective
cultural contexts.

Despite fraught and often bloody settler-kangaroo cultural histories, kangaroos now hold a special place in the contemporary Australian national imaginary.
Because of the quirkiness of their upright bodies and the way that they hop on two
legs, these fascinating marsupials are widely regarded as the quintessential and
emblematic national animal. They have come to stand for Australia. This prevailing attitude to kangaroos no doubt pre-disposed the children in our Canberra study
to recognise them as 'special', to assume their 'natural' belonging, and to perhaps
also underpinned their efforts to get to know the kangaroo bodies as well as they
possibly could. Possums are not quite as renowned, but they are well adapted to
urban environments and many Australians live in close proximity to them. They
are commonly referred to as a nuisance, but also as harmless and cute enough to
be tolerated and enjoyed. This kind of attitude was reflected in the Canberra early
childhood education centre, where the resident possums caused additional maintenance work, but were nevertheless still very welcome.

In Canada, deer are also central to settler national identity, but this is framed
around their association with wilderness and hunting culture, rather than their idiosyncratic physical attributes. The increasing presence of deer in the leafier cities,
such as Victoria, British Columbia, is largely interpreted as this species being out of
place, causing inconvenience, and becoming a problem. The children and mule deer
in the Victoria, British Columbia, ethnographic study forged convivial relationships
against the backdrop of this kind of rising (human) concern, enabled by their mutual
curiosity and the deer's apparent docility. It was not so easy for the children in the
Vancouver study to get close to the racoons. Throughout Canada, there is scant public sympathy for raccoons. Their reputation as cunning tricksters, and potentially
aggressive intruders, who pose a health and safety risk to all humans, but especially
children, overrides their cute, comedic appeal – at least to those adults who are
responsible for children. Both children and educators in this study had to work hard
against the prevailing culture of fear and protection from 'dangerous' racoons, to
build some kind of uneasy proximate relationship, based upon the children's and
racoons' mutual curiosities, interests, and identifications. In all of these cases, the
ethical possibilities of children's relations with deer, kangaroos, possums, and raccoons have been contingent upon these various existing local ways of 'knowing'
about them – but have also exceeded them in various degrees.

The conditions of ethical possibility surrounding the children's relations with ants and worms are quite different. While it may be relatively easy for us humans to emotionally engage with other charismatic mammals, it is much harder for us to identify with small insects and invertebrates, who have no visible faces and bear little resemblance to us. In Chapter 3, we described how it was the worms' and ants' ubiquitous and manifold presence, the allure of their complex micro-worlds, and their incessant earth-moving activities that captured the children's curiosity, and drew them into relationship. Because this relationship was not based upon any form of self-recognition, or reduction to human prototype, it afforded a different kind of ethics – one that Myra Hird (2010, 2012, 2013) has variously referred to as an 'inhuman' or 'environmental' ethics of mutual vulnerability.[1] Again, the contingencies and local specificities of the Canadian and Australian ethnographic studies, as well as the creatures themselves, differentiated the means and extent to which this ethics was realised.

In the more structured pedagogical program at the Victoria early childhood education centre that eventually centred around the children's hands-on sensory engagement with the worms' wet wriggling bodies in the classroom vermicompost bin, there was a stronger tendency to revert to the more familiar (humanist) ethics of human care for the other. In spite of the children's embodied encounters with worms as potent soil-makers, it was easier for the educators to emphasise and the children to realise the worms' vulnerability at their hands, than vice versa. The feisty ants offered a different set of affordances and contingencies to the children in the Canberra early childhood centre. The ants were not nearly as enticing to handle, and their capacity to swarm fast and bite hard was a clear retaliation when the children's clumsy feet, poking sticks, and slapping hands threatened their lives. In other words, the mutual vulnerability of children and ants was very evident.

It matters that we are both situated in ex-British colonies. In both locales, our studies of child–animal common world relations have been framed by the messy cultural politics and dire ecological legacies of settler-colonised lands – in particular the mass species extinctions that are so closely linked to colonisation. Nowhere is this more evident than in Chapter 4, where we have focused upon the ways in which eco-nationalist cultural texts increasingly position settler Australian and Canadian children and young people as stewards or saviours of endangered native animals. As we noted, the (predominantly urban) targeted child audiences are unlikely to ever meet the native animals featured in these texts – the endangered Australian bilby and the rare British Columbian spirit bear – as the few that remain live in very remote areas. So the children's relationships to these native animals rely entirely upon the ways in which the texts interpolate them. In this time of anthropogenic extinctions, the goodwill and ethical underpinnings of such cultural texts, which champion the need to save threatened native animals from extinction, might seem very worthy. However, we have drawn upon indigenous settler-colonial critique (Langton 2012; Hall 2015) and critical feminist Anthropocene scholarship (Haraway *et al.* 2015) to problematise settler stewardship's unwitting reiteration of nature/culture binaries, its anthropocentric conceits and heroisms, and its indigenous appropriations and erasures. Drawing upon Deborah

Bird Rose's (2004, 2011) work, we have counter-offered a decolonising ethics of ecological reconciliation, which recognises that all lives and deaths are entangled in the web of exchange that makes life on earth possible and promotes recon-ciliation and recuperation through firming up connections, rather than rehearsing further separations. The emphasis of this ethics is that ecological reconciliation is something we all (human and nonhuman, indigenous and non-indigenous) need to do together.

Similarly, the *where* of our studies also mattered a great deal in Chapter 6. Here, we focused on indigenous human-animal cosmoecologies in two of the harshest environments on earth – the Artic regions of Canada and central desert region of Australia. Needless to say, we held fast in our determination not to celebrate and promote individual (white, human) altruism and largesse as the only way to 'save' the planet, and focused instead on what we might learn from tundra and desert indigenous child-dog hunting relations about a recuperative ethics of kinship obligation. The Inuit, Arrernte, and Warlpiri notions of child-dog kin-ship with which we engaged have absolutely nothing to do with the sentimental Euro-western tendency to regard pet dogs as 'part of the family' – a notion that inevitably requires the dogs to be infantilised, anthropomorphised, and assimi-lated into human schema. These indigenous concepts are based upon resolutely pragmatic and unsentimental understandings that reciprocal interspecies kinship responsibilities and obligations are essential to survival. They derive from undi-vided inter-corporeal ontologies that simply presume all living creatures are kin, because we are all part of the same recycling matter, and we all come from the land (Turner 2010).

The ethical proposition of a recuperative ethics of kinship obligation, was formed by bringing Inuit, Warlpiri, and Arrernte narratives of child-dog kin-ship into conversation with Vinciane Despret and Michel Meuret's (2016, p. 27) reparative notion of an 'ecology of obligations', whereby cohabiting humans and other animals share the common purpose of seeking new 'ways of inhabiting a world that is being destroyed while resisting, locally and actively, this destruc-tion' (Despret and Meuret 2016, p. 30). In other words, by combining some of the wisdom of local indigenous eco-cosmologies that have survived for millennia with contemporary multispecies thinking about how we might survive the current planetary ecological crisis together, we have come up with a recuperative ethics of kinship obligation. It is premised on a shared understanding that all earthlings are mortal kin, in the broadest sense of the term, and thus obligated and beholden to each other for our mutual survival.

Scale and significance

Throughout this book we have confronted the question of scale and significance. The biggest 'time-space-mattering' (Barad 2011) notion of all – the paradoxical figure of the Anthropocene – and all that it portends for the future of life on earth as we know it, hovers over our discussions of the common worlds of children and animals, and the ethical implications of their entangled lives, including the minu-tiae of their daily encounters. So does the pervasive figure of British colonisation.

We have continually witnessed how settler colonial cultural and ecological legacies and displacements are played out in the local 'contact zones' (Haraway 2008, pp. 316–317) of child–animal relations in our situated Canadian and Australian studies. In many of the chapters, we have traced patterns of global/imperial/national forces, circuits, and exchanges that converge in the 'throwntogetherness' of children and animals on 'extraverted' local grounds (Massey 2005), and shape the conditions of possibility for their relations. This entire book is interwoven with such articulations of scale and significance.

There is also another way in which we have confronted the question of scale and significance. Although we take the gravitas of precarious ecological futures completely on board, and this awareness fortifies our commitment to intergenerational and interspecies environmental justice, we completely reject the grandiosity of human-centric responses to the predicaments we face. None of the ethical responses we have explored are about the largess of superior and heroic humans coming to the rescue and saving the world. For the important task of learning how to recuperate, re-figure, and reconstitute damaged worlds, we have deliberately cast our lot with the seemingly insignificant and minor players who are usually discounted in the high stakes business of world changing. We have refocused the world-changing spotlight on the small, modest, and incremental interspecies achievements of children and animals, which take place (usually un-noticed) on the grounds of their local common worlds. We have done this in order to learn more about how these common worlds are already being remade together in the ordinary, everyday encounters, interactions, and emerging relations between unlikely partners (Haraway 2008, 2016).

In honing the spotlight onto small but nevertheless significant local child–animal re-worlding events, and considering their ethical affordances, we have not intended that they be taken up as blueprints that can be scaled up and replicated elsewhere. As Anna Tsing (2012, 2015) points out in her damning critique of scalability as a conventional measure of research significance, scaling up requires uniformity. And yet the emergent differences and specificities of the local child–animal relations we have studied defy uniformity. The most significant feature of these relations is that they are productive and transformative. They make a difference. They do not reproduce the same. For us, the point of tracing multiscalar threads through children's and animals' lives, as well as zeroing in on the ethical affordances of specific local child–animal common world relations, has not been to develop a grand theory that can generalise from the particular to the universal (or vice versa), but to better understand the entanglement of life, and from there to search for fresh and hopeful ways of reconceptualising and recuperating our damaged worlds together. The significance of our study lies in the art of noticing and learning from small interspecies re-worldling achievements, one by one, not in scaling up.

Re-envisaging hopeful common world futures

In this book, we have tackled what we believe to be the most prescient challenge of our times – the intensifying precarity of life on earth as we know it. We have taken this challenge as a timely moment to interrupt the business as usual of humanist scholarship and to promote a paradigm shift. For us, this shift has involved three basic moves. Firstly, it has led us to contemplate the implications of this ecological challenge not just for children, but for the entangled lives and futures of children and animals. Secondly, it has propelled us to consider how ordinary, everyday child–animal common world relations might illuminate some ethical ways forward without dragging us back into the dangerous presumption of human exceptionalism. And, finally, it has redoubled our efforts to re-envisage hopeful common world futures.

To recap in the briefest (albeit rather derivative) terms, our hopeful re-envisaging, or speculative imagining, goes something like this:

In these common world futures, *multispecies belonging*[2] is an unnegotiable baseline. This is because we are all seeking refuge in our damaged, mixed up, and species-depleted common worlds, and we need each other. Knowing this helps us to persist with the discomfort and challenge of *learning how to live with incommensurable differences* at close quarters. For many of us humans, this kind of learning requires us to *continually reassess our place and agency in the world, as just one amongst many actors*.[3] With our sights firmly set on collective interests and actions, we get on with the job of *recuperating our damaged common worlds together to make them more liveable for all*.[4] We realise the danger of over-predicting what kind and extent of recuperation might be possible. To avoid relapsing into human-centric grandiosity, we aim for realistic *modest, small, incremental changes, or partial recuperations*,[5] not all-encompassing solutions. Recognising that there is no final or ultimate resolution we resolve to persevere, or to *stay with the trouble*,[6] and we keep reminding ourselves, by way of encouragement, that *worlds are continually made and remade through relations*,[7] no matter how seemingly insignificant.

This entire book is the product of such re-envisaging. We have composed it in 'bag lady' style (Le Guin 1996) by throwing all sorts of joyful ingredients into our carrier bags. It is full of the best ideas from the best scholars we know, our reflections upon select child–animal relations that we witnessed during our ethnographic fieldwork, our analysis of some interesting children's animal texts, and some of the wisdom we have gleaned from the 'otherwise' ontologies of indigenous peoples in our respective home countries.

Out of this carrier bag book we extract hope for the future common worlds of children and animals – those in our studies and beyond.

Notes

1 Even when humanist ethics are extended to nonhuman animals – because they are clever like us, or they look something like us – they still rely upon Levinas' (1991) notion that ethics is the act of recognising ourselves in the other through face-to-face encounter. When there is no face-to-face encounter, as in the case of our relations with the microbial lifeworld, Hird (2010, 2012, 2013) argues that we require an ethics that does not default to human prototype.

2 Our appreciation of the need for multispecies belonging in a time of extinctions has been sharpened by the work of Deborah Bird Rose (2011) and Thom van Dooren (2014). See also van Dooren and Rose (2012).

3 All of the scholars we follow take (more than human) collective agency as axiomatic to the ways in which world works. However, Bruno Latour (1987) and Michel Callon (1986) were among the first theorists to attribute agency to networks of human-nonhumans actors.
4 Donna Haraway has been talking about finding ways to make worlds more livable for years. In her latest book (Haraway 2016), she notes that the current ecological crisis brings compelling new urgency to this task.
5 Again, it is Haraway (2008) that reminds us that the most we can hope for is partial recuperation. There is no final solution.
6 We cannot halt the earth's cascading systems collapses, the best that we can do, according to Haraway (2016), is to 'stay with the trouble'.
7 Haraway is certainly not alone in insisting that relations are productive, but she has coined the terms 'worlding' and 're-worlding' to indicate that it is the mortally entangled relations between species that both sustains and constantly re-calibrates life on earth (Haraway 2008, 2016).

References

Anderson, K., 1995. Culture and nature at the Adelaide Zoo: At the frontiers of 'human' geography. *Transactions of the Institute of British Geographers*, 20(3), 275–294.

Barad, K., 2011. Nature's queer performativity. *Qui Parle: Critical Humanities and Social Sciences*, 19(2), 121–158.

Callon, M., 1986. Some elements of a sociology of translation: Domestication of the scallops and fishermen of St. Brieuc Bay. *In*: J. Law, ed. *Power, action and belief: A new sociology of knowledge?* London: Routledge and Kegan Paul, 196–223.

Despret, V. and Meuret, M., 2016. Cosmoecological sheep and the arts of living on a damaged planet. *Environmental Humanities*, 8(1), 24–36. doi:10.1215/22011919-3527704.

Gibson, K., Rose, D.B. and Fincher, R., eds., 2015. *Manifesto for living in the Anthropocene*. Brooklyn: Punctum.

Hall, L., 2015. My mother's garden: Aesthetics, indigenous renewal, and creativity. *In*: H. Davis and E. Turpin, eds. *Art in the Anthropocene: Encounters among aesthetics, politics, environments and epistemologies*. London: Open Humanities Press, 283–292.

Haraway, D., 2008. *When species meet*. Minneapolis: University of Minnesota Press.

———., 2016. *Staying with the trouble: Making kin in the Cthulucene*. Durham: Duke University Press.

Haraway, D., Ishikawa, N., Gilbert, S., Olwig, K.R., Tsing, A.L. and Bubandt, N., 2015. Anthropologists are talking – About the Anthropocene. *Ethnos*, 81(3), 535–564.

Hird, M.J., 2010. Meeting with the microcosmos. *Environment and Planning D: Society and Space*, 28, 36–39.

———., 2012. Animal, all too animal: Toward an ethic of vulnerability. *In*: A. Gross and A. Vallely, eds. *Animal others and the human imagination*. New York: Columbia University Press, 331–348.

———., 2013. Waste, landfills, and an environmental ethics of vulnerability. *Ethics and the Environment*, 18(1), 105–124.

Langton, M., 2012. *The conceit of wilderness ideology*. Boyer Lectures (Lecture 4), Australian Broadcasting Commission. Available from: www.abc.net.au/radionational/programs/boyerlectures/2012-boyer-lectures-234/4409022 [Accessed 9 February 2018].

Latour, B., 1987. *Science in action: How to follow scientists and engineers through society*. Cambridge, MA: Harvard University Press.

Le Guin, U.K., 1996. The carrier bag theory of fiction. *In*: C. Glotfelty and H. Fromm, eds. *The ecocriticism reader: Landmarks in literacy ecology*. Athens: University of Georgia Press, 149–154.

Levinas, E., 1991. *Totality and infinity*. A. Lingis, trans. Norwell: Kluwer Academic Publishers.

Lorimer, J., 2010. Moving image methodologies for more-than-human geographies. *Cultural Geographies*, 17(2), 237–258.

———., 2015. *Wildlife in the Anthropocene: Conservation after nature*. Minneapolis: University of Minnesota Press.

Massey, D., 2005. *For space*. London: SAGE.

Povinelli, E., 2016. *Geontologies: A requiem to late liberalism*. Durham: Duke University Press.

Puig de la Bellacasa, M., 2010. Ethical doings in naturecultures. *Ethics, Place, and Environment: A Journal of Philosophy and Geography*, 13(2), 151–169.

Rose, D.B., 2004. *Reports from a wild country: Ethics for decolonisation*. Sydney: UNSW Press.

———., 2011. *Wild dog dreaming: Love and extinction*. Charlottesville: University of Virginia Press.

———., 2012. Multispecies knots of ethical time. *Environmental Philosophy*, 9(1), 127–140.

———., 2016. 'About love and extinction', *Deborah bird rose love and extinction blog*. Available from: http://deborahbirdrose.com/about/ [Accessed 27 February 2018].

Rose, D.B., van Dooren, T., Chrulew, M., Cooke, S., Kearnes, M. and O'Gorman, E., 2012. Thinking through the environment: Unsettling the humanities. *Environmental Humanities*, 1, 1–5.

Steffen, W., Crutzen, P. and McNeill, J.R., 2007. The Anthropocene: Are humans now overwhelming the great forces of nature? *Ambio: A Journal of the Human Environment*, 36(8), 614–621.

Stengers, I., 2016. *In catastrophic times: Resisting the coming barbarism*. London: Open University Press.

Tsing, A.L., 2012. On nonscalability: The living world is not amenable to precision-nested scales. *Common Knowledge*, 18(3), 505–524.

———., 2015. *The mushroom at the end of the world: On the possibility of life in capitalist ruins*. Princeton: Princeton University Press.

Turner, M.K., 2010. *Iwenhe tyerrtye: What it means to be an Aboriginal person*. Alice Springs, Australia: IAD Press.

van Dooren, T., 2014. *Flight ways: Life and loss at the edge of extinction*. New York: Columbia University Press.

van Dooren, T. and Rose, D.B., 2012. Storied places in a multispecies city. *Humanimalia: A Journal of Human/Animal Interface Studies*, 3(2), 1–27.

Index

Printed in the United States
by Baker & Taylor Publisher Services